David Brown

Rapid, Low Cost Modeling and Simulation

Integrating Bayesian and Neural Networks as an Alternative to an Equation-Based Approach

VDM Verlag Dr. Müller

Impressum/Imprint (nur für Deutschland/ only for Germany)
Bibliografische Information der Deutschen Nationalbibliothek: Die Deutsche Nationalbibliothek verzeichnet diese Publikation in der Deutschen Nationalbibliografie; detaillierte bibliografische Daten sind im Internet über http://dnb.d-nb.de abrufbar.
Alle in diesem Buch genannten Marken und Produktnamen unterliegen warenzeichen-, marken- oder patentrechtlichem Schutz bzw. sind Warenzeichen oder eingetragene Warenzeichen der jeweiligen Inhaber. Die Wiedergabe von Marken, Produktnamen, Gebrauchsnamen, Handelsnamen, Warenbezeichnungen u.s.w. in diesem Werk berechtigt auch ohne besondere Kennzeichnung nicht zu der Annahme, dass solche Namen im Sinne der Warenzeichen- und Markenschutzgesetzgebung als frei zu betrachten wären und daher von jedermann benutzt werden dürften.

Coverbild: www.purestockx.com

Verlag: VDM Verlag Dr. Müller Aktiengesellschaft & Co. KG
Dudweiler Landstr. 125 a, 66123 Saarbrücken, Deutschland
Telefon +49 681 9100-698, Telefax +49 681 9100-988, Email: info@vdm-verlag.de
Zugl.: Fairfax, George Mason University, Diss., 2004

Herstellung in Deutschland:
Schaltungsdienst Lange o.H.G., Zehrensdorfer Str. 11, D-12277 Berlin
Books on Demand GmbH, Gutenbergring 53, D-22848 Norderstedt
Reha GmbH, Dudweiler Landstr. 99, D- 66123 Saarbrücken
ISBN: 978-3-639-07965-4

Imprint (only for USA, GB)
Bibliographic information published by the Deutsche Nationalbibliothek: The Deutsche Nationalbibliothek lists this publication in the Deutsche Nationalbibliografie; detailed bibliographic data are available in the Internet at http://dnb.d-nb.de.
Any brand names and product names mentioned in this book are subject to trademark, brand or patent protection and are trademarks or registered trademarks of their respective holders. The use of brand names, product names, common names, trade names, product descriptions etc. even without
a particular marking in this works is in no way to be construed to mean that such names may be regarded as unrestricted in respect of trademark and brand protection legislation and could thus be used by anyone.

Cover image: www.purestockx.com

Publisher:
VDM Verlag Dr. Müller Aktiengesellschaft & Co. KG
Dudweiler Landstr. 125 a, 66123 Saarbrücken, Germany
Phone +49 681 9100-698, Fax +49 681 9100-988, Email: info@vdm-verlag.de

Produced in USA and UK by:
Lightning Source Inc., 1246 Heil Quaker Blvd., La Vergne, TN 37086, USA
Lightning Source UK Ltd., Chapter House, Pitfield, Kiln Farm, Milton Keynes, MK11 3LW, GB
BookSurge, 7290 B. Investment Drive, North Charleston, SC 29418, USA
ISBN: 978-3-639-07965-4

David Brown

Rapid, Low Cost Modeling and Simulation

DEDICATION

In loving memory of my father, Joseph P. Brown,
who taught me you are never too old to go back to school.

ACKNOWLEDGEMENTS

I wish to offer my deepest and most sincere gratitude to my advisor, Dr. Kathryn Blackmond Laskey, who was willing to take on the advisor role mid way through my Ph.D. process. Her countless hours and encouragement were instrumental in helping me complete this work. Also deserving my gratitude is my committee chair, Dr. Andrew Sage. I am honored to have such a well known and respected man in the field of Systems Engineering as my committee chairman. I also wish to express my gratitude and recognize my two other committee members, Dr. David Shum and Dr. Zoran Duric for their help and encouragement. Additionally, I would like to recognize and thank my first advisor, Dr. Dennis Buede, who helped me develop the ideas that were the foundation of this research. Other people who deserve recognition for assistance in this research include Vernon Gordon and Michael Hocker of the U.S. Naval Test Pilot School for providing the infrared receiver and radar test data, Brent Boerlage of Norsys Software Corp. for providing the beta version of the Netica API enabling the creation of continuous variables, and fellow student Ning Xu of George Mason University for providing her discretization software allowing a comparative analysis of different discretization methods.

TABLE OF CONTENTS

1. Introduction

1.1 Modeling and Simulation as a Systems Engineering Tool

Modeling and Simulation (M&S) is an important tool for performing trade studies in systems engineering. M&S provides designers with the ability to examine a large number of virtual designs before constructing a prototype or system. This provides a variety of benefits, including balancing requirements with available funding and schedule, determining risk areas, building efficient test plans and reducing test requirements. The aggressive use of modeling and simulation is one of the few systems engineering tools that have demonstrated the simultaneous achievement of a better product brought to market in less time at a lower cost [DTSE&E, 1996]. M&S should be used throughout the life cycle of a system. A successful M&S program begins with an aggressive effort early in concept exploration and definition. During this period of product development, M&S may be the only data available upon which to make decisions concerning cost and performance trade-offs as well as other engineering trades. However, this currently requires an extensive up-front investment. Discussions with program managers and studies have shown that M&S is not extensively used in smaller programs because they do not have the resources to make this investment. Reducing the cost of modeling and simulation so that it becomes affordable for use in smaller programs and product developments would represent a substantial advancement to the field of systems engineering. Additional challenges for systems engineering are the increasing complexity of systems and the increasing pressure for interoperability between systems. This system-of-systems concept means that almost every system is a subsystem in some larger system. The M&S required to support a system-of-systems concept typically requires the integration of large, complex models. Additionally, there is a drive towards reuse of existing models to reduce cost. Although reuse of M&S can save considerable time and money, integration of models built by different companies or organizations that were never designed to communicate with each other can pose significant technical challenges.

Five areas have been identified that currently inhibit integration of different M&S tools [DTSE&E, 1996]. They are

- availability of data descriptions

- security/sensitivity of data
- lack of equation-based understanding of complex phenomena
- hardware and software limitations
- variable resolution

There are ongoing efforts to define data standards. Efforts are also being made to better secure data so that only those who are authorized can access it. Information technology continues to grow at a breakneck pace while software capabilities expand to match advances in hardware capability. This research focuses on the remaining two technical areas: modeling of phenomena for which equation-based models do not exist and integration of models at multiple levels of resolution. The research will also investigate reducing the high cost and length of time required to build models using current equation-based modeling techniques.

1.2 Terms and Definitions

For the purpose of this research, the term model is defined as a virtual representation of a system, entity, phenomenon or process [Acquisition Functional Working Group, 1999]. Models are simplified representations of systems at a particular point in time intended to promote understanding of a real system [Bellinger, 2002]. Models are typically specified as parameterized families. Setting the parameters to different values allows the user to examine the behavior of a particular simplified system within the family. Often a modeler varies parameters in a process called sensitivity analysis to examine their impact on model outputs [Morgan et al., 1990]. Models of systems that evolve in time are often executed in computer simulations where the computer representation consists of a sequence of representations of the system at a succession of time points. The most common examples are training simulators that contain a model of a physical system such as a car or aircraft. In many simulations, a human operator can make inputs to the model that cause it to change state over time. Thus, a person flying an aircraft simulator is currently at a specific point in the synthetic sky at a specific set of parameters such as altitude, airspeed, attitude, etc. because of control inputs made in past time. Inputs made at time present specifies either deterministically or probabilistically where the airplane is at any future time. The time within a simulation can either be accelerated or slowed down to better understand interactions. Although modeling and simulation are two different things, they share many common attributes and problems. When describing issues relevant to both models and

2

simulations, this group of virtual representations is referred to as Modeling and Simulation (M&S).

There are different types of models and simulations. Most models currently in use are equation-based. An equation-based model uses mathematical expressions to describe the real world it attempts to replicate in a virtual environment. Thus, the first requirement to construct an equation-based model is to obtain or develop a set of equations that describes to the best of our ability how we believe things in the real world work. Another type of model is a joint probability distribution, typically expressed as a Bayesian network [Jensen, 1996]. A Bayesian network represents a set of related *random variables*, which represent properties or features of the world whose values are unknown, but for which probabilistic information is available. Each possible configuration of values of the random variables represents a possible way the world might be. Probabilities are assigned to these configurations by specifying a local probability distribution for each random variable given a small set of random variables that directly influences it. Model inputs are represented as evidence for the random variables corresponding to the inputs. A Bayesian network inference algorithm can be used to make probabilistic predictions for output random variables given the input random variables. The data required to specify the local probability distributions can be obtained empirically from measurement by recording the number of times something occurred when specific conditions were observed, or by eliciting the probabilities from an expert.

Two other terms used to describe all models and simulations are fidelity and resolution. Fidelity, defined from the Latin roots as faithful, is the level of accuracy with which the virtual world of M&S actually represents (or is faithful to) the real world [Acquisition Functional Working Group, 1999]. With equation-based models for well-understood phenomena, the level of fidelity generally represents how many influencing variables were taken into account when creating the model. For poorly understood phenomena, it may be impossible to construct models with high fidelity. Determining how many variables to include and the level of fidelity necessary for the intended use are major issues in construction of equation-based models. Resolution is the degree of precision represented by a model or simulation [Davis & Ziegler, 2000]. A model that calculates an answer in inches provides more resolution than one providing an answer in feet. Although fidelity and resolution are different concepts, they are related in that they must be matched to each other. Specifying an output to several decimal places in resolution is of little use if the model does not have the fidelity necessary to make the extra decimal places meaningful. Alternately, if a quantity is calculated to the nearest tenth of an inch but the output is rounded to the nearest foot, the fidelity is lost. Thus the resolution of the output should match the fidelity with which the output was generated.

1.3 Benefits and Costs of Modeling and Simulation

Modeling and Simulation is an analysis and control tool in the field of systems engineering. As computer power continues to increase, model builders are able to build increasingly more complex and accurate virtual representations of real-world entities [Zittel, 1998]. These have provided impressive improvements in product quality, reductions in time to develop products and lower product costs. Table 1-1 provides a summary of these improvements from a study of the use of modeling and simulation in both the public and private sectors.

Table 1-1[1]

Measured Benefits of Modeling and Simulation

Who	What	Traditional Method	New Method with M&S
TRW	Radar Warning System Design	96 man-months	46 man-months
TARDEC	BFV Engineering Analysis	4-6 man-months	0.5 man-months
TARDEC	Low Silhouette Tank Design	55 engineers – 3 years	14 engineers – 16 months

---

[1] DTSE&E study (1996) "Study on the Effectiveness of Modeling and Simulation in the Weapon System Acquisition Process", Final Report.

General Electric	Engine Fan Blade	4 weeks	A few hours
Lockheed Martin	Engineering Mock-ups	2100 hours	900 hors
Lockheed Martin	Changes per Final drawing	4	2
Lockheed Martin	Physical Mock-ups	$30M each	None
Lockheed Martin	Design Verification	Baseline	30% - 50% reduction from baseline
IBM	Computers	10,000 parts 4 years	4000 parts 2 years
Motorola	Cellular devices	Baseline	50% reduction in product cycle time
Sikorsky Aircraft	Helicopter External Working Drawings	38 draftsmen 6 months	1 engineer 1 month
NAVSEA	Ship Seakeeping Analysis	27 days	3.5 days
NAVSEA	Radar Cross Section Analysis	57 days	17 days
Comanche Helicopter Program	Source Selection	Prototype Fly-off $500M	Simulator/Surrogate Aircraft Fly-off $20M

As can be seen from these examples, most of the success stories found during this study involved large government programs and/or products developed by large corporations.

Although the benefits of Modeling and Simulation have been demonstrated, these benefits come at a steep price. An aggressive M&S effort requires an extensive up front investment. Development of the Boeing 777, a recognized business success case in which M&S played a significant role, required an up front investment of roughly one hundred million dollars [Garcia, Gocke & Johnson, 1994]. The M&S core body of knowledge states under limitations that "M&S

tools are not generally inexpensive and require an up-front investment cost" [Acquisition Functional Working Group, 1999]. This statement is backed up by results of a study looking at the cost of the M&S effort on Department of Defense programs summarized in table 1-2.

Table 1-2[2]

Department of Defense M&S Cost Data

Program	Approximate Total Program Cost	M&S Expenditures to date	% of M&S with Cost Data
LPD-17	$10B	$38M	100
ATACMS/BAT	$5B	$25.2M	100
Javelin	$4B	$48M	100
AN/BSY-2	$3B	$58.3M	100
SADARM	$3B	$14.6M	78
V-22	$37B	$50.2M	44
FAAD C2	$1B	$37.6M	40

This same study chartered by the Department of Defense found that program managers do not consider DoD-wide M&S investments as either cost or schedule effective [Hicks & Associates, Inc., 2001]. Additionally, small programs simply do not have the resources to make this up-front investment [DTSE&E, 1996]. One study found that if resources permitted greater use of M&S, small development projects could achieve improvements that were comparable to the gains achieved by the larger programs in Table 1-1 [Brown, 1999].

If the cost and time to develop M&S tools can be reduced, this would represent a large potential improvement in both government and commercial product developments. In the United States, 50% of the gross domestic product is generated by small businesses.[3] Improvements on the scale of those shown in table 1-1 represent an enormous untapped potential in product development if modeling and simulation can be done more quickly and at lower cost.

---

[2] Hicks & Associates, Inc., (2001) "Modeling and Simulation Survey Briefing".

1.4 Why is M&S so Expensive?

The high cost and time associated with building equation-based models, even for products the equations of which are well understood, can be seen by looking at the modeling of a simple system. The particular case is the building a model for simulating the performance of a spring-powered car [Brown, 1999]. The model was constructed for use by students taking a course in systems engineering at the Defense Systems Management College and was designed to demonstrate the value of modeling and simulation when trade-offs are made during the design phase. The object was to conduct a series of trades to find a combination of variables that provided good performance for only two performance requirements at the lowest cost. The final equation of motion for this simple vehicle had 34 variables and 8 coefficients. Modeling and simulation of complex systems may require an extremely large number of variables and coefficients as well as the equations that relate them together. It is highly unlikely that a company making spring-powered cars could afford even a tiny fraction of the costs in table 1-2. The study in the case then compared a group of students who used the model with a control group that did not have access to the model. The use of M&S in the design phase resulted in better performance at lower cost for the same amount of time spent on the project [Brown, 1999]. The benefits of using M&S in the design phase of any project, regardless of size, are significant.

Once any model is built, it must be verified and validated before use [Acquisition Functional Working Group, 1999]. Verification tests that the model has been implemented correctly, while validation checks to see that the model or simulation accurately represents the real world system. To correctly validate a model, the actual system is tested over the range of values that the model or simulation will be used to predict. Any critical areas such as the edges of the operating envelope should be included in the validation test data. The model predictions are checked against the test data. If the model does not agree within specified limits in any area with the test data, further tests are conducted to determine the cause of the difference. These causes are then incorporated into the model and the results checked again. This process continues until an acceptable agreement between the test data and the model is obtained. This explains the finding in the M&S Core Body of Knowledge which states that attempts to create high-fidelity models rapidly drive up the cost of a modeling effort [Acquisition Functional Working Group, 1999]. The key to wider use of modeling and simulation in product

---

[3] Data from the Small Business Administration website.

development, particularly for smaller projects, is to significantly reduce the time and expense of current methods.

1.5 Probability Models

The proposed alternative to be explored in this research is to model systems and subsystems as probability distributions instead of deterministic equations. Probability distributions can consistently represent different resolutions as they contain not only a central tendency, but also a measure of certainty. Thus a low-resolution model can be represented by a distribution with a larger spread to account for the lack of detail in the model. A high-resolution model of the same entity would have the same central tendency but less spread in its distribution representing finer detail and more certainty. Probability distributions can be combined through subsystem interaction while maintaining consistency that is captured by the central tendency and spread of the final system level distribution.

The proposed tool to model probability distributions is a Bayesian network. Bayesian networks encode a complete and coherent probability distribution over many variables and can be used to evaluate both causal and evidential influences. The conditional probabilities can be either entered directly or learned from a data set. Bayesian networks can be manually constructed by creating nodes that represent the variables in a problem and creating arcs representing cause and effect relationships between the variables. Some advanced Bayesian network packages can learn network structure directly from data sets. Thus, Bayesian networks can model phenomena that can be measured or observed but for which equations are not known. The research will explore the use of Bayesian networks for rapid generation of probabilistic models of systems and subsystems.

To effectively use Bayesian networks to model systems and subsystems, they must be able to integrate with existing equation-based models. This integration requires two-way data exchange between the Bayesian and equation-based subsystem elements. Since Bayesian networks are most frequently used in artificial intelligence applications, there is the potential for other benefits to the science of modeling and simulation beyond the central research. Complex systems frequently have humans that must interact with the system. This interaction influences the outcome of dynamic simulations. Utility and decision nodes can also be added to Bayesian networks resulting in influence diagrams capable of making decisions. Utility nodes allow the assignment of value to various states of input nodes while decision nodes allow assignment of

8

decision options. An influence diagram is capable of making probabilistic decisions using utility theory. Thus, it should be possible for the equation-based portion of an integrated simulation to generate input data to set the node states of a Bayesian network or influence diagram. The network could then make a probabilistic inference or optimized decision that could be sent back to the equation-based simulation and would then affect the outcome. By soliciting conditional probabilities and utility values from people who would control the system in operation, it should be possible to have the influence diagram represent a user or group or users within a computer simulation. This would allow trade studies in which human decision making and human interaction are simulated without the need to have humans present.

Human decision making is currently implemented in many war game simulations. A study of implementation finds the majority of systems use a rule-based approach to decision making. The study found that this method is seriously lacking in realism because the decision process was too stereotypical, predictable, rigid and doctrine limited. Rule-based programming requires either complete information to make a decision or an even larger number of rules to determine how to handle missing information. The same study looked at three areas for improving the representation of human decision making within simulations. These areas were rank-dependent utility, multi-attribute utility and game theory. Rank-dependent utility uses non-linear transformations from objective probabilities to decision weights. Multi-attribute utility models the process by which multiple, and often conflicting objectives are traded off against each other to make a decision. Game theory models systems in which multiple interacting intelligent agents with conflicting objectives make decisions based on their beliefs about the environment and behavior of other agents [Pew and Mavor, 1998]. The ability of Bayesian networks and influence diagrams to learn from data sets could also be used for system control within a simulation. The probabilistic relations between nodes can be learned from the output of a simulation. Using utility nodes to define system goals, an influence diagram can learn the parameters that resulted in the most favorable outcomes. By structuring the simulation so that a Bayesian network or influence diagram could make decisions that affect the simulation outcome, the influence diagram could act as a feedback control system by learning to make decisions that result in outcomes most favorable to the goals. Even if the conditions of the problem undergo adaptive change, the network can continue to learn from the outcomes keeping it current. By contrast, rule-based models must be reprogrammed with new rules in response to a changing

9

environment. One German automobile manufacturer is successfully reducing its costs by using a large influence diagram to control its production using real time updates of sales data to predict the number of parts to be delivered and to forecast the delivery schedule [Lerat, 2002]. Bayesian networks are also being integrated with a number of software applications. Intel has released a set of Bayesian network software libraries to assist software developers in building machine learning capabilities into their software applications [Knight, 2003]. Microsoft has used Bayesian networks for years in their Office Assistant and online troubleshooting wizards [Breese and Koller, 1997]. One of the most recent applications of Bayesian networks is in filtering Spam Email [Sahami et al, 1998].

1.6 Research Hypotheses

The objective of the research is to assess empirically how the probabilistic modeling approach compares to equation-based models of the same phenomena. The two areas of comparison are the time required to construct each model or simulation and the accuracy with which the model or simulation predicts test data of the physical system that is virtually represented. The systems to be tested are listed in table 1-3.

Table 1-3

Evaluation Model Matrix

Appendix	NAME
A	Amplifier
B	LRC circuit
C	Wing
D	Elevator control
E	Radar
F	Forward Looking Infrared (FLIR)
G	Aircraft Radar Cross Section
H	Loan Application
I	Car Electrical Repair
J	Home Heating System

K	Commuter
L	Air Defense
M	Robotic Vehicle

The models and simulations are each described in a separate appendix as listed in table 1-3. The research hypotheses are as follows:

Hypothesis #1
Null hypothesis: Bayesian networks have the same average percent error as equation-based models.
Alternate hypothesis: Bayesian networks do not have the same average percent error as equation-based models.

Hypothesis #2
Null hypothesis: Bayesian networks require less than or equal time to construct compared with equation-based models.
Alternate hypothesis: Bayesian networks require greater time to construct compared with equation-based models.

Each of these hypotheses will be tested against three types of Bayesian network models:

- Manually constructed networks using human judgment for probabilities
- Manually constructed networks using formulae for probabilities
- Computer constructed networks using structural and probabilistic learning algorithms

The goal of the research is to demonstrate that Bayesian networks can be constructed more quickly and that these networks will have approximately the same accuracy as an equation-based model of the same system or process. Of the three methods, the computer generated networks are expected to have the fastest times of construction. The research will demonstrate that Bayesian networks and equations-based models can be integrated together and operate within the same modeling or simulation environment. Within this integrated environment, the research will show that Bayesian networks or influence diagrams can make decisions which can be used to control a simulation.

2. Current Methods of Modeling and Simulation

2.1 Methods of Handling Variability

As described in chapter 1, equation-based modeling and simulation is currently performed by identifying the relevant dependent and independent variables that characterize the behavior of the system and then developing equations that describe how they relate to each other. Input values are identified and the model or simulation is executed to provide an output. This can be demonstrated by looking at a simulation of a home heating system shown in figure 2-1.

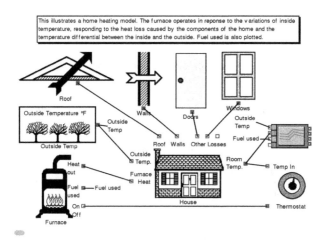

Figure 2-1

Home Heating Model[4]

A thermostat controls the inside temperature of the house. The furnace adds heat when it is on, while heat is lost through the roof, walls, doors and windows. The amount of heat loss is also dependent on the outside air temperature. A single run output over a 24 hour period using handbook insulation values and average high and low temperatures for the month of January in the Washington DC area is shown in figure 2-2.

---

[4] Model provided with the Extend M&S software package. Original author is unknown.

Value   Y2

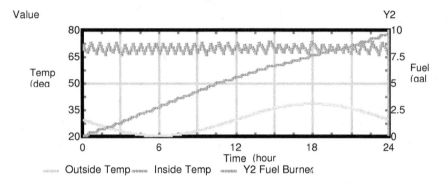

Figure 2-2

Home Heating Simulation Output

As the model is run over a 24 hour period, one can see the variation in outside air temperature shown as the sine function in the lower portion of the chart. The inside air temperature is the saw tooth function at the top and the fuel oil consumed is the step function going from lower left to upper right. Using handbook values and average temperatures, one would expect this house to use 9.54 gallons of fuel each 24 hours.

If one were to go to the house shown above and measure the fuel used each 24 hours, it is highly unlikely one would get a value of precisely 9.54 gallons on any given day. This is because the output of any system can be heavily dependent on variability in the inputs, unmodeled factors and random noise. It is difficult to control variables outside a laboratory environment. In the home heating model, one can control to some extent the variables such as the insulation in the roof, number of doors, windows, etc. However, one has no control over the outside air temperature. If variation cannot be controlled, the current practice in M&S is to run a sensitivity analysis on the variables [Clemen, 1996]. A sensitivity analysis varies the independent variables over their expected range of values in the anticipated operational environment to determine the sensitivities (or gradients) with respect to the dependent variables of interest [Arsham, 2002]. The model or simulation is run multiple times with the variable on which the analysis is being performed being incremented by a fixed amount on each run. The analysis begins at either the highest or the lowest value of the sensitivity range and continues

until the opposite end of the range is reached. A sensitivity analysis of seven variables along
with the single solution using point values is provided in figure 2-3.

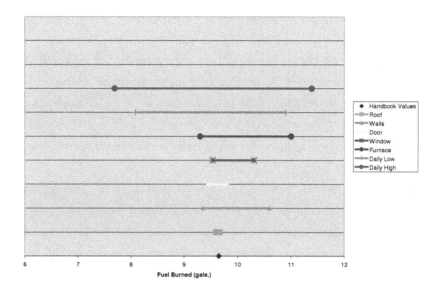

Figure 2-3

Sensitivity Analysis for Home Heating Model

After the analysis is run, one can ascertain how much impact changes in the independent
variable have on dependent variables of interest. As can be seen in figure 2-3, the example
model exhibits a low sensitivity to variability in the heat loss through the roof over the expected
range of the heat loss variable. The model exhibits the highest sensitivity to outside air
temperature. In this example, the high and low temperatures are two variables within the model
which cannot be controlled.

The current process of M&S sensitivity analysis is flawed when viewed from the
perspective of probability theory. If the dependent variables are continuous then there is a zero
probability of getting the exact answer provided by the model or simulation on any one run
[Larson, 1969]. This is a highly likely in any engineering problem. A sensitivity analysis
provides a more complete answer by specifying a range over which the answer may lie.

14

However, this answer is incomplete in that it provides no information about where within the range the answer is most likely to fall. The vertical difference between the lines of figure 2-3 is provided only for separation to distinguish between the different variables plotted on the same chart for comparison. There are no units on Y-axis of figure 2-3. It also is deficient in the variables are varied one at a time while holding all others constant. This analysis is sufficient in estimating model sensitivities only of the effects of the parameters on the model are independent and monotonic [Bankes, 1993]. No probability or confidence can be attached to the range of even the most sensitive variable. Furthermore, variables that show little sensitivity when varied independently may exhibit strong sensitivity when varied in combination with other variables. Thus, running a sensitivity analysis may not capture the true range of the solution space of the dependent variables.

A review of research literature finds different approaches have been developed to address the limitations described above. Some new modeling packages now allow a Monte Carlo sampling to be performed using a distribution function instead of fixed incremental steps. By running the model or simulation multiple times and plotting a histogram of the outputs, one can now determine not only the range of possible outcomes but also which of these answers are more or less likely to occur. An example of this procedure for the daily low temperature is provided in figure 2-4. The analysis is done for 1000 runs using a normal distribution of temperature with a mean of 20 degrees F and standard deviation of 6 degrees F.

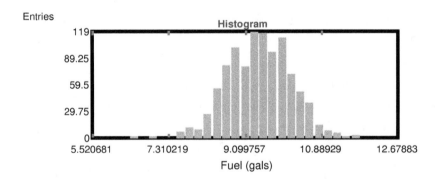

Figure 2-4

Monte Carlo Sampling of Daily Low Temperature

15

One can see from the example of figure 2-4 that the most likely output values lie within the center of the range. Because this method is evaluating one variable at a time, it assumes independence between the variables and does not capture correlation between variables.

Another proposed method is to do a pair-wise comparison of all variables with each other to determine correlation [Clemen and Really, 1999]. This method would identify situations in which two variables are correlated such that one variable either positively or negatively impacts the other. It also identifies if synergistic effects are present where two variables acting in combination may result in a dramatically different result than either variable separately. If either condition exists, a sensitivity analysis is run varying both variables simultaneously. An example of a dual sensitivity analysis for the daily high and low temperature is provided in figure 2-5.

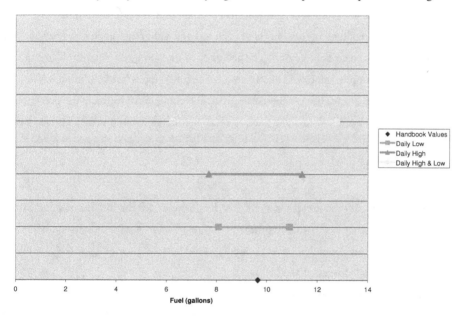

Figure 2-5

Dual Sensitivity Analysis of Temperatures

For comparison, the individual sensitivities and handbook value solution are also shown. As can be seen from figure 2-5, varying the daily high and low simultaneously from the low end of each variable range to the high end results in a new sensitivity range. This new range is 76% greater than the greatest individual analysis range. This method provides a more accurate measure of the possible range of outputs. However, it captures only pairs of correlated variables. If three or more variables are highly correlated as a group, this will not be detected and the actual range of responses may be understated. Since this method uses incremental steps, it does not provide any indication of where within this expanded range the answer is more or less likely to lie.

The most complete method found was the modeling of all variables that exhibit variation as random variables [Morgan et al., 1990] [Bankes, 1993]. Each random variable is set to a distribution function that describes how the variable behaves. A Monte Carlo method is then used by sampling from each distribution over multiple computer runs to obtain a probability distribution of the dependent variables. This provides a complete probabilistic solution to the problem in that all correlations and synergistic effects are captured, the complete range of possible outputs is captured, and the likelihood of each answer within the range is specified. The Monte Carlo method has linear complexity. The accuracy of the output distribution is dependent only on the number of samples generated and is not dependent on the number of inputs. The Monte Carlo method also allows use of standard statistical techniques to estimate the precision of the output distribution. This allows a small sample set to be run and from this data an estimate of the number of total samples required to achieve a desired confidence can be calculated. This can be done using the formula

$$m > (2 * c * s / w)^2$$

where m is the total number of samples required, c is the deviation for the unit normal enclosing the probability (distance enclosed by the two-tailed confidence interval of a unit normal), s is the standard deviation from the sample and w is the width of the desired interval enclosed by the two tailed confidence interval [Morgan et al., 1990].

To illustrate Monte Carlo sampling, the home heating model is again used as an example. A distribution is first fit to each of the variables. An example of the loss factor for the doors is shown in figure 2-6. The handbook value for door losses for doors with weather stripping is 18. New weather stripping would likely have a normal distribution about the handbook value. The loss factor for installed weather stripping will only increase with time as the weather stripping and seals crack and wear out from use. The door loss distribution for an average door is therefore assessed as a gamma distribution with most of the values falling to the right of the handbook value. Similar distributions are then set for the roof, walls, and windows. The fuel

17

distribution for a Monte Carlo simulation with 1000 runs with probability distributions for losses from the roof, walls, doors and windows, the furnace efficiency and the daily high and low temperatures is shown in figure 2-7.

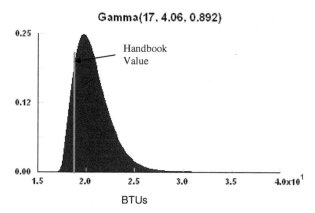

Figure 2-6

Door Loss Distribution

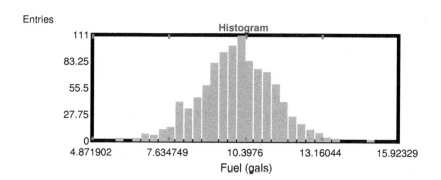

Figure 2-7

Monte Carlo Simulation with Probability Distribution of Inputs

18

The Monte Carlo simulation results in a distribution that is approximately Gaussian for fuel consumed with a mean of 10.25 gallons and a standard deviation of 1.29 gallons. It is important to note that the mean of the distribution is higher than the 9.54 gallon answer generated using handbook values. This demonstrates the importance of the distribution assigned to the variables. In this case, the sensitivity analysis showed that the loss was likely to become worse over time due to age and wear causing higher average fuel consumption in the Monte Carlo simulation. Some companies such as caterpillar tractor currently use this method in the design of their products [Blood, 2001].

Although this method provides a complete solution, it does so at a high computational cost. The Monte Carlo method is considered to be a crude or brute force technique because of the large number of samples and computational workload to achieve the output distribution [Morgan et al., 1990]. Figure 2-8 illustrates computation times for the same model using each of the above methods for assessing variability of results.

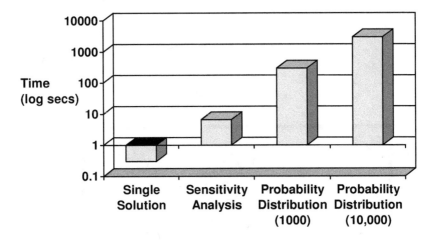

Figure 2-8

Computer Execution Times for Home Heating Model

Figure 2-8 is plotted using a logarithmic scale because of the large differences in execution times. A single run of this model on a 600 MHz notebook computer takes 0.3 seconds. Running either a single or dual sensitivity analysis increases the execution time to 6.6 seconds. However, running a Monte Carlo simulation for either a single or multiple variables increases the execution time to 5 minutes for 1000 runs and to 50 minutes for 10,000 runs. There is a trade-off using this method between accuracy and speed in that generating more runs improves the accuracy of the output distribution while increasing the run time to generate the distributions. Although there was little difference between the means and standard deviations between 1000 and 10,000 runs, the maximum fuel consumption increased from 17.2 to 19.9 gallons. Although this may not be significant for a heating system trade study, a maximum possible value that is 16% higher could be critical for a safety related trade study.

2.2 Multi-Fidelity and Multi-Resolution Modeling

One trend in the field of M&S is the increased integration of models and simulations. As discussed above, cost considerations have led to increasing reuse of existing M&S tools in new developments [Hollenbach, 2001] [Konwin, 2001]. There is also an increased development of system of system concepts in which systems are integrated together to create larger, more complex systems. This requires the integration of different models and simulations that were not designed to interoperate with each other [Doyle, 2000]. These models were constructed with fidelities and resolutions as specified for their original purpose. When these different models or simulations are integrated, the result is mixed fidelity and mixed resolution. An analogy to this problem is to calculate the area of a room with one person measuring the width with a tape measurer and giving you the result to the nearest sixteenth of an inch and the other pacing off the length and reporting the result in feet. The two measurements have very different accuracies with respect to the actual lengths and were reported at different resolutions. The difficulty then lies in how to combine these two inputs to estimate the area.

A review of literature and phone interviews with experts in this area shows that the current approach to multi-fidelity model integration is through interaction at the lowest common level [Hollenback, 2000], [Smith, 1999]. In the above analogy, the width measurement would be rounded to the nearest foot for the calculation of the area and the output would be specified in square feet. The drawback to this approach is that the accuracy from the high fidelity width measurement was completely lost. The resulting answer would only be accurate to plus or minus a square foot.

What is proposed is a probabilistic solution to engineering problems in cases where there is mixed fidelity and/or resolution in the input parameters. Using the earlier example of finding the area of a room with two measurements of fidelity and resolution, assume that the length and width are two different model elements of different fidelity and resolution which must be combined. Let us further assume that the length is reported at 125 3/16 inches and the width at 9 feet. Using current methods, one would convert the length to 10.5 feet and multiply it by 9 feet to get an answer of 94.5 square feet. Using a probabilistic method one would first quantify the accuracy of the measurements. Let us assume that the length is normally distributed with a mean of 125.1875 inches and standard deviation of .08 inches while the width is normally distributed with a mean of 9 feet and standard deviation of 1 foot. Modeling the length and width as probability distributions and calculating the area is shown in the model in figure 2-9.

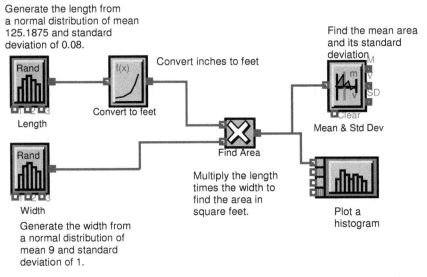

Figure 2-9

Area Calculation Model

Using this method, the length is converted to feet so that both inputs are in consistent units. Running ten thousand Monte Carlo samples provides an output distribution of the dependent variable area as shown below in figure 2-10.

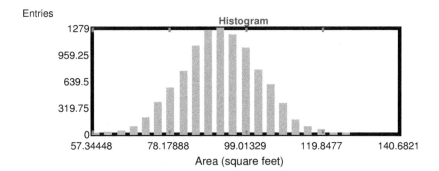

Figure 2-10

Area Calculation Model Output

Since both distributions were normal, the mean value of 93.99 square feet is very near the previous answer of 94.5 square feet. However, figure 2-10 shows that the actual answer could take on a wide range of values. The additional information from a probability distribution as an output provides valuable information in engineering applications. In the above example, if one is measuring the area to cover the floor with some type of protective coating, then 108 square feet of coverage should be purchased to assure a 90% probability of having enough coating to cover the entire floor.

3. Bayesian Networks and Influence Diagrams

3.1 Network Description

Bayesian networks are directed graphs for representing probabilistic dependencies among variables [Jensen, 1996]. Bayesian networks encode a complete and coherent probability distribution over many variables and can be used to evaluate both causal and evidential influences. A Bayesian network consists of a directed acyclic graph that represents dependencies among variables, together with local probability distributions defined for small clusters of directly related variables. Directed acyclic graphs consisting of nodes which represent the variables and arcs (or directed edges) that describe cause and effect relationships or statistical associations between the variables [Jensen, 1996]. Each variable has a finite set of mutually exclusive states. The graph may contain no directed cycles, or paths that lead from a node to itself and follow the direction of the arcs. Each node is assumed to be conditionally independent of its non-descendents given its parents. An example Bayesian network is shown in figure 3-1.

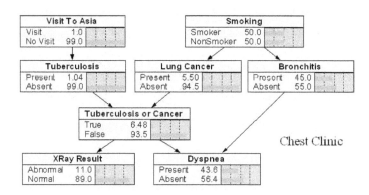

Figure 3-1

Example Bayesian Network

Root nodes have no incoming arcs (e.g., nodes "Visit to Asia" and "Smoking"). Child nodes have one or more incoming arcs.

Probability information in a Bayesian network is specified via a local distribution for each node. The local distribution for a root node is simply an assignment of a probability to each

23

state such that the probabilities sum to 1. The local distribution for a child node is given by a conditional probability table. This table shows the probability of each possible state of the child conditional on each possible state of all of its parents. Thus a child node with x states and having y parents also with x states would have a conditional probability table with $x^{(y+1)}$ entries. The joint distribution for all variables in the network is given by the product of the local distributions for all the nodes:

$$P(X_1,\ldots,X_n) = \prod_{i=1}^{n} P(X_i \mid \underline{X}_{pa(i)})$$

where $\underline{X}_{pa(i)}$ denotes the parents of variable X_i.

The conditional probability that variable E takes on value e given that H takes on value h is defined by the equation:

$$P(E=e|H=h) = P(E=e \text{ and } H=h)$$

A straightforward consequence of this definition is <u>Bayes Rule</u>, a powerful mathematical relationship by which probabilities can be modified to incorporate new evidence:

$$P(H|E) = P(H) * P(E|H) / P(E)$$

The first term, $P(H|E)$ is referred to as the "posterior probability" or the probability of H given evidence E. The term $P(H)$ is the prior probability of H. The term $P(E|H)$ is the "likelihood" and gives the probability of the evidence assuming hypothesis H is true. The last term is the probability of E that acts as a normalizing or scaling factor [Niedermayer, 1998]. Bayes Rule can also be written in odds likelihood expressed as:

$$\frac{P(H=h_1|E)}{P(H=h_2|E)} = \frac{P(H=h_1)}{P(H=h_2)} \frac{P(E|H=h_1)}{P(E|H=h_2)}$$

The second factor is called the likelihood ratio for h_1 versus h_2. This form clearly demonstrates that evidence increases the probability of h_1 relative to h_2 if and only if the evidence is more likely under h_1 than h_2.

This can be demonstrated by entering evidence into figure 3-1. States "No Visit", "Smoker", "Abnormal" and "Present" are entered into the network as shown in figure 3-2.

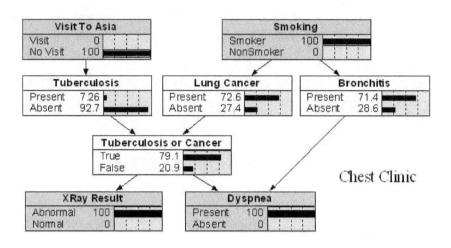

Figure 3-2

Chest Clinic XRay Abnormal

This evidence leads to beliefs of a low probability of Tuberculosis with moderate probabilities of Lung Cancer and Bronchitis. If node "XRay Result" is changed from "Abnormal" to "Normal", the network of figure 3-3 is obtained.

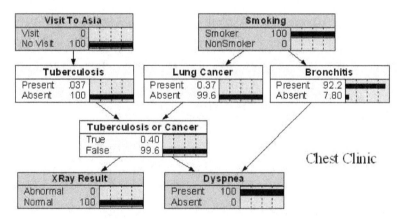

Figure 3-3

Chest Clinic Normal XRay

This single change of evidence results in a new set of beliefs. There is now a very low probability of Tuberculosis or Cancer, and a high probability of Bronchitis.

There are three types of nodes that are used in networks: natural, utility and decision nodes. Bayesian networks contain only natural nodes, while influence diagrams contain all three types. Natural nodes are defined by the probability relations of parents and children described above. Natural nodes can be of two types: discrete and continuous. Discrete nodes have a finite or countable number of states typically designated by a list of state names for finite state nodes or indexed by integers for countable state nodes. Continuous nodes can take on values of real numbers in a range between a lower and upper limit. This range of numbers can be either a continuous function or discretized into bins. If discretized, the range is divided into a countable number of mutually exclusive, yet continuous bins. The bin width (or difference between the lower and upper limit of each bin) does not have to be the same size for each bin.

When utility and decision nodes are used in a directed graph representing options, probabilities over consequences or utilities over consequences in a decision problem, this is referred to as an influence diagram. Influence diagrams extend Bayesian networks to model optimal decision making along with uncertainty about the state of the world. Natural nodes represent aspects of the world. Utility nodes allow the assignment of values to each combination of states of its parent nodes. Decision nodes have states that represent actions. These actions can be either intervening actions which cause a state change of variables in the model or non-intervening where there is no impact on the network model [Jensen, 1996]. Influence diagrams optimize decisions options by summing the probabilities of the parents of the utility node times the utility values of each decision option. The highest value represents the optimum decision under the given inputs.

3.2 Bayesian Networks as Engineering Models

Bayesian networks and influence diagrams have several features that have the potential to improve deficiencies identified with equation based models. A Bayesian network models the relationship between nodes as probability distributions. Thus, if a probabilistic relationship is efficiently computable, a Bayesian network can calculate a probability distribution for a dependent variable in a single computation cycle where as a Monte Carlo simulation requires many cycles with corresponding high times of computation to calculate the same distribution. Bayesian networks also can calculate an exact answer to a probabilistic query. The local distributions for a Bayesian network can also be specified in more than one way providing additional flexibility. One method is to represent relationships between variables as a formula. The formula can have random components, allowing the expression of parameterized probability distributions. Thus, the equation-based formula approach to model construction can be used if desired when defining relationships between nodes. Probability distributions can also be specified by eliciting the relations from experts. These predictions include the type of distribution and central tendency. The certainty the expert assigns to the prediction will be reflected in the spread of the distribution. The relations can also come from test data or observations. It is also possible to use a combination of empirical data and informed engineering judgment in representing a model [Laskey, 2002].

Another advantage of Bayesian networks is their ability to reason under uncertainty [Laskey, 2002], or in engineering terms, to generate a plausible model output even when some of the inputs are missing. An equation-based model must have all the inputs to solve the equations that drive it. Bayesian networks can reason from the top down or bottom up [Murphy, 2000]. In top down reasoning, the causes are entered and the effects are calculated by the network. In bottoms up reasoning, the effects are entered and the network will provide the most likely causes. One can therefore run the model as a normal engineering application where the inputs are entered and the model provides the outputs. The models can also be run in the reverse direction where the outputs are entered into the model and the most probable states of the inputs are determined.

An example Bayesian network for the area of a room is shown below in figure 3-4. In this example, the cause and effect relationships are shown by the direction of the arcs. Input variables "Length" and "Width" are continuous chance nodes that are inputs to a formula in the deterministic continuous node "Area". In this example, "Length" is defined as a normal distribution of mean 13.0 and standard deviation of 1.0 feet. "Width" is defined as a uniform distribution of minimum 8 and maximum of 12 feet. "Length" has a resolution of one foot while "Width" has a resolution of 0.5 feet. The distributions and resolutions were chosen to demonstrate the ability to handle different distributions in a multi-resolution environment. Node "Area" is defined by the formula "Length" times "Width". When the network is compiled, the distribution of node "Area" is calculated as shown in figure 3-4. One can enter values for "Length", "Width" or both values resulting in a probability distribution for "Area".

Additionally, one can also enter a value for "Length" and "Area" and the model will provide the probability distribution for "Width". Such flexibility allows model builders to better understand the results, how the model works and how it might be improved [Laskey, 2002]. Equation-based models cannot provide solutions to problems with missing data without building a complex set of rules for handling missing inputs. They also can not work in the opposite direction unless the entire model is reconstructed.

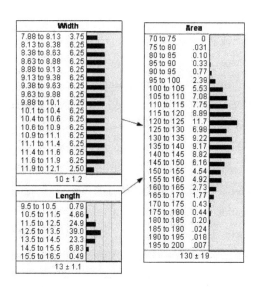

Figure 3-4

Area Probability Model

Bayesian networks can be constructed using multiple methods. They can be built manually using a graphical interface to generate the nodes, define the states (discrete nodes) or bin widths (continuous nodes) and draw the arcs between nodes. Probabilities for each node may then be entered directly into the node probability tables or generated from an entered formula. Bayesian networks can also be constructed using computer algorithms. Network structures can be generated by computer algorithms which learn the relationships between variables from data

28

sets. Some new Bayesian network packages such as Hugin, MSBN, Bayesian Network PowerSoft and Kevin Murphy's MATLAB Toolbox are capable of learning structure from data sets. Computer methods also exist for determining the dividing points for bins if discretization of continuous data is required. Most Bayesian network software packages contain the capability to learn the local probability tables from files containing cases generated from the distribution to be learned. This eliminates the need to input these numbers manually when data are available. Research has also been done in the area of learning structure and probabilistic relations with missing variables, incomplete data, or both simultaneously [Friedman, 1998].

3.3 Limitations of Bayesian Networks

Bayesian network theory, inference and software are mature with respect to handling of discrete variables. Theory and software are not nearly as mature with respect to continuous variables. Although there are methods for learning structure directly from continuous data, studies have shown that discretizing the data prior to learning the structure results in both higher efficiency with respect to learning and greater accuracy of the learned network (Monti and Cooper, 1997). A number of different methods of discretization have been proposed. These methods can be broken down further into the following classes: supervised vs. unsupervised, dynamic vs. static, global vs. local, splitting (top-down) vs. merging (bottom-up), direct vs. incremental and univariate vs. multivariate [Liu et al, 2002].

Supervised methods of discretization use class information while unsupervised methods do not. If the network structure is not known, then an unsupervised method must be used. Previous research has shown that supervised methods result in better inference and accuracy than unsupervised methods [Liu et al, 2002] [Xu, 2003]. Dynamic methods use an iterative approach such that information from previous calculations is used in calculating the discretization. Static methods discretize in a single operation. Dynamic methods can take considerably longer to execute due to the iterative nature of the approach. Global methods discretize data based on the entire structure of the network. Local methods discretize each variable individually in isolation. For large networks, local methods will take much longer to execute since each variable is discretized individually. Splitting methods take the existing intervals of a variable and divide it into two new intervals. Merging methods do the opposite by combining two existing intervals into a single one. Direct methods divide a continuous variable range into a number of bins simultaneously. Incremental methods start with a simple discretization and iteratively attempt to improve the discretization using a scoring technique. Incremental methods are time consuming due to their iterative nature. Univariate methods discretize each continuous node one at a time. Multivariate methods discretize multiple nodes at the same time. Multivariate methods take less time as they are more efficient.

Most commercial software packages use unsupervised methods to discretize continuous data prior to structural learning. The two most frequently used methods of unsupervised discretization are the equal width and equal frequency methods. The equal width method sorts the data, takes the distance between the minimum and maximum and sets the bin widths at equal intervals between the minimum and maximum based on the number of bins. The equal frequency method sorts the data, counts the total number of data points and then divides up the bin widths so that an equal number of points are placed in each bin. If the data points are not divisible into a whole number, the last bin may contain more or less data than the other bins. Although unsupervised methods may be adequate for learning of structural relationships between variables, they are inadequate for defining the bins of continuous nodes for learning of probabilistic relations from data sets. An example of the performance of two models created using these two unsupervised methods using a 10 bin discretization is shown in figure 3-5. Although either method provides a reasonable prediction in areas were the data is relatively constant (Time > 1.20), both methods provide poor results in areas of rapid change (0.00 – 0.80). Because of this limitation, networks created using uninformed discretization methods are unsuitable for use as the basis for comparison to equation-based models in this research.

A number of supervised methods have been proposed for determining the bin dividing points in continuous nodes once the structure has been determined. These include entropy methods such as ID3, D2, MDLP, Contrast and Mantaras Distance. They also include merging methods such as ChiMerge, Chi2, and ConMerge as well as other methods such as 1R, Marginal Entropy, Zeta and Adaptive Quantizer (Liu et. al., 2002). Two other iterative methods of discretization are the Hill Climbing [Pearl, 1985] and Markov Chain Monte Carlo (MCMC) [Gilkes et al., 1996]. All these methods look at discretizing a continuous variable by either splitting or merging the cut points based on iterative refinement of individual variables.

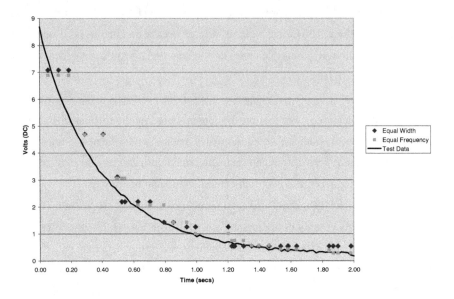

Figure 3-5

Unsupervised Discretization Examples

A multivariate discretization method using a latent variable model is another proposed method for improving discretization [Monti, 1999]. This method discretizes continuous data prior to structural learning by using the latent variable to score potential relationships between variables and then discretizing on the higher scores. Although this method demonstrated improved structural learning as compared to unsupervised methods and direct learning from continuous data, it is computationally complex and would appear to be infeasible for very large databases. The number of possible latent variable scores would increase exponentially with the number of variables using this approach. Like the merge and split methods, the latent variable model is designed for discretization of continuous data for construction of Bayesian network classification models.

The approach taken in this research is that if the structure is not known, the data is discretized twice. If the structure is unknown, an unsupervised method must be used for the first discretization so that a structural learning algorithm can be used. However, once the structure is determined, continuous nodes should be discretized again using a supervised method to improve performance in the final network, as research has shown that supervised methods provide better results as compared to unsupervised methods [Liu et al, 2002] [Xu, 2003]. Considering the different methods available, a preferred approach to the final discretization to maximize accuracy while minimizing computational time would be a supervised, static, global, direct, multivariate approach.

Another limitation of current Bayesian network software packages is that the networks they produce by learning from data sets can not generate a solution for a different set of discrete conditions not in the data set. A network constructed from a data set with input resistor values of 1000Ω and 2800Ω would not be able to predict outputs with an input of 1800Ω. Similarly, a Bayesian network model can not interpolate between two states. If training data were provided at intervals of one tenth of a second, the model could not interpolate an answer at fifteen hundredths of a second. If the output variable is continuous, the model would provide an answer of whatever bin that value fell within. This is a severe limitation in that for engineering applications, a network could not produce an answer for which it did not have training data unless a formula or expert prediction were used to create the probability table.

3.4 Conclusions and Discussion

Recent advances in Bayesian network research provide some exciting possibilities for modeling and simulation of systems. It may be possible to model systems which can be observed but for which the mathematics, relations and all variables are not available. Combining structural learning, automatic bin calculation for continuous variables and probabilistic relation learning, it would be possible to create Bayesian network models through computer algorithms with minimal human input. These attributes are expected to save large amounts of time in constructing a model as compared to equation-based methods. Verifying that these attributes exist and quantifying them for both manually constructed and computer constructed models is the principal focus of this research. The two principal obstacles that must be addressed to achieve this capability are an efficient, accurate method of discretization of continuous nodes once the structure is known and developing a method to allow a network to make inferences to conditions not contained in the input data set.

4. The Derivative Method of Continuous Variable Discretization

4.1 Derivative Algorithm

Data sets for engineering models typically contain multiple continuous variables and mixtures of discrete and continuous variables. A discretization technique for engineering models must be multivariate with the capability to discretize all related continuous variables at the same time. A computationally efficient method that can be directly calculated from the data set is proposed in this research. To illustrate this method, a typical discretization problem for two related continuous variables is shown in figure 4-1.

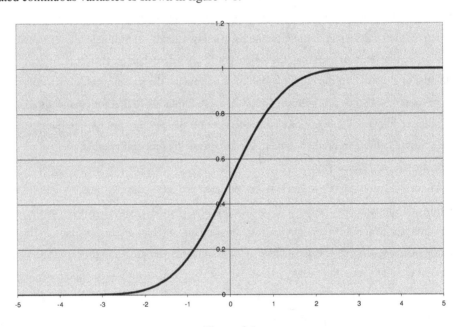

Figure 4-1

Example Discretization Problem

In this example there are two continuous variables with the parent node data shown on the X-axis and child on the Y-axis. A manual discretization of these variables would place the bin dividing points far apart in areas were the curve is flat (-5 to -2, 2 to 5), and closer together where the curve is steep (-2 to 2).

The derivative method is based on the concept of spacing cut points with respect to how much the variables change. There are two versions of the algorithm: one in which the nodes to be discretized have continuous parents or children described in 4.2.1 and one in which a continuous node has only discrete parents and/or children described in 4.2.2. The derivative measures the change with one variable with respect to another. In the example above, calculating the derivative of Y with respect to X would result in a relatively small number in flat areas of the above curve and a much larger number in steep areas. The derivative method measures the change across the entire data set by numerical derivatives of the function and setting bin boundaries so that approximately equal amounts of change fall into each of the bins. Given that variable X contains values $x_1, x_2, ..., x_n$ and the variable Y contains values $y_1, y_2,...,$ y_n such that X and Y have an equal number of points, the derivative at each point is approximately equal to

$$\Delta y_i / \Delta x = (y_{i+1} - y_{i-1}) / (x_{i+1} - x_{i-1})$$

for i = 2 to n-1 and

$$\Delta y_i / \Delta x = (y_i - y_{i-1}) / (x_i - x_{i-1})$$

for i = n.

At i = 1 the first cut point is established. All change is then calculated in reference to this starting point up to and including the final point at i = n.

The total amount of change is

$$\text{total change} = \sum_{i=2}^{n} | \Delta y_i / \Delta x |.$$

The absolute value is required to obtain a measurement of the total amount of change irrespective of the direction of change in the data set. The amount of change to place in each bin is defined by

$$\text{change per bin} = \text{total change} / \# \text{ bins}.$$

The final step is to find the dividing points that define the bin widths. This is done by summing the absolute values of the derivative at each point. Each time a derivative is added the sum is checked to see if it has exceeded the change per bin limit. If it has not exceeded the value the next absolute derivative value is added. If the sum exceeds the limit there is a cut point in the interval between the current and last value of x and y. The cut points for x and y are a percentage of the difference between the values of the variables before and after the limit was exceeded. If j is the integer count of the number of derivative values that were added when the sum exceeded the change per bin value, then the percentage is

$$\text{percent} = (\text{change per bin} - (\text{current sum} - \Delta y_j / \Delta x)) / (\Delta y_j / \Delta x).$$

The cut points in data sets X and Y are then found by

$$xcut = x_{j-1} + \text{percent} * (x_j - x_{j-1})$$

and

$$ycut = y_{j-1} + \text{percent} * (y_j - y_{j-1}).$$

The value of the current sum must be reset to the amount in excess of the change per bin value:

$$\text{current sum} = \text{current sum} - \text{change per bin}.$$

The process is now repeated until all the cut points are found.

Figures 4-2 and 4-3 show the results of applying the derivative method discretization process for five and eight bins respectively.

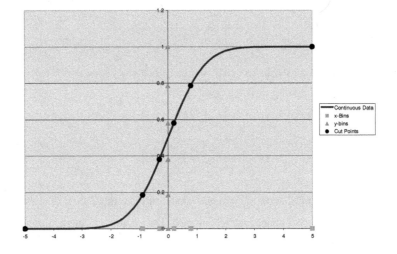

Figure 4-2

Five-Bin Discretization

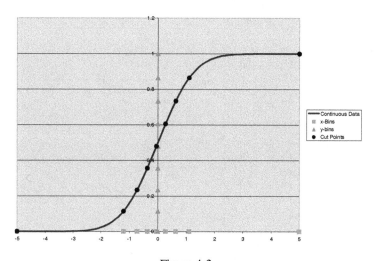

Figure 4-3

Eight-Bin Discretization

The derivative method is straightforward for this simple case of two variables. The more difficult challenge is extending the method to all possible combinations of discrete and continuous parents and children. The derivative method needs only two sets of rules to handle all cases: one rule set for continuous variables with either continuous parents or children and a second set for continuous variables with only discrete parents and children.

4.1.1 Continuous Nodes with Continuous Parents or Children

The derivative method simultaneously discretizes all continuous nodes that are related to each other through arcs with other continuous nodes. An example network is provided in figure 4-4. The first step is to identify all continuous node chains within a network. In the example of figure 4-4, nodes A, D, E, F, G and H are continuous nodes while B and C are discrete nodes. There are two continuous chains in the example: A-D-G and E-F-H. Note that although B is a parent of a node in each chain, it is discrete so that it separates the continuous groups into two separate chains. In this example, each chain is discretized separately.

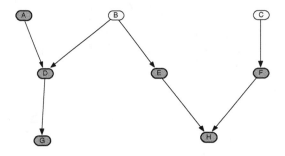

Figure 4-4

Example Mixed-Node Network

Once all the nodes in a continuous chain are identified, the next step is to locate a root chain node. A root chain node has no continuous parents. In the example network, A, E and F are root chain nodes. A root chain node may or may not be a root node of the network. It is only necessary to identify one root chain node, even though more than one may exist. When taking a derivative, the operation is conducted on the dependent variable with respect to the independent variable. A root chain node is selected as the X variable because it is independent of all other nodes in a continuous chain.

Before the derivatives are calculated, the data set must be placed in the correct format. The data for all continuous nodes in the chain is sorted on the chosen root chain node from the lowest to highest numeric value. Duplicate data points are combined by averaging the data for each duplicate value of the root chain node. This operation serves two purposes. First, it ensures that the derivative process does not break down due to an infinite slope. If a data set contains multiple points for the same value of the root chain node X, then the difference in X between data sets would be zero and the derivative $\Delta Y / \Delta X$ would be infinite. Second, it allows the process to ignore the influence of discrete parents and children with arcs to or from the continuous variables in the chain. If the slopes are similar, this can be seen in the example of figure 4-5.

A Bayesian network model of the system of figure 4-5 contains a continuous variable (volts) with three parents (L, R and C) that were sampled as discrete variables and one continuous parent (time). Each of the discrete variables has two states resulting in eight total curves. One variable, L, has no influence on the system, resulting in four distinct curves. The derivative method places an equal amount of change in each bin. This could be done by measuring the derivatives of all eight curves individually and by adding these individual derivatives until a cut point is found. This is accomplished using the formula

$$\text{total change} = \sum_{i=1}^{m} \sum_{j=1}^{n} \sum_{k=2}^{o} | \Delta y_{ijk}/\Delta x |$$

where m is the number of all continuous nodes in the chain other than the chosen root chain node, n is the number of rows in the CPT table for each continuous node of m and o is the

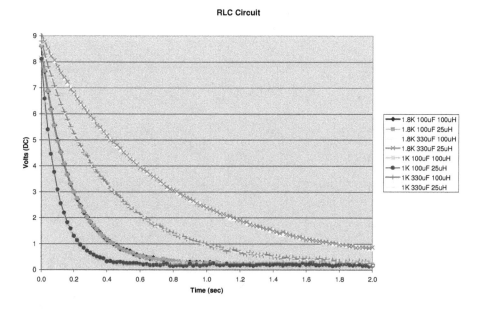

Figure 4-5

Example Data Set

38

number of rows in the data set (number of unique values for the root chain node). For complex chains with multiple parents, this presents a lengthy calculation. If the derivatives are all positive or negative values, an alternate method allowing the discrete parents to be ignored can be used, greatly simplifying the problem. This is accomplished by using the average value of the dependent variables (all continuous curves in the chain) at each point of the independent variable (root chain node). This results in an alternate equation of the form

$$\text{total change} = \sum_{k=2}^{o} | \Delta((\sum_{i=1}^{m} \sum_{j=1}^{n} y_{ij}/ \Sigma n_m) /\Delta x)_k |$$

where the derivative is now calculated at the average value of all the continuous variable curves. An example of this is shown in figure 4-6 by finding the average of the four distinct curves of figure 4-5. For this example, $o = 101$, $m = 1$ and $n = 4$ and $\Sigma n_m = 4$.

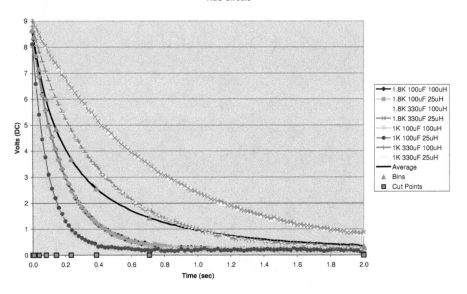

Figure 4-6

Average Curve Example

One can observe from figure 4-6 that the cut points for the average of the four distinct curves are the same as if one calculated the derivative for each curve individually and then found the cut points by summing all four derivatives at each point in time until the cut threshold is exceeded. This can also be demonstrated mathematically. The change threshold defining a cut point using the individual curves is

$$\text{threshold} = \sum_{i=1}^{m} \sum_{j=1}^{n} \sum_{k=2}^{o} |\, \Delta y_{ijk}/\Delta x \,| \; / \; \# \text{ bins}$$

while the threshold using the average of the curves is

$$\text{threshold} = \sum_{k=2}^{o} |\, \Delta((\sum_{i=1}^{m} \sum_{j=1}^{n} y_{ij}/ \Sigma n_m)/\Delta x)_k \,| \; / \# \text{ bins}$$

To define a cut point, the algorithm sums the derivatives until the threshold value is exceeded. For the case where all derivatives are evaluated and having summed x points of the independent variable, a check is made to see if the threshold has been exceeded using the inequality

$$\sum_{k=2}^{x} \sum_{i=1}^{m} \sum_{j=1}^{n} |\, \Delta y_{ijk}/\Delta x \,| \;\leq\; \sum_{i=1}^{m} \sum_{j=1}^{n} \sum_{k=2}^{o} |\, \Delta y_{ijk}/\Delta x \,| \; / \; \# \text{ bins}.$$

If at k=x the inequality is true, the next value of k is added until the inequality is false. At that point a cut point is calculated as previously described in section 4.1. For the case where the average of the derivatives is evaluated and having summed x points of the independent variable, a check is made to see if the threshold has been exceeded using the inequality

$$\sum_{k=2}^{x} |\, \Delta((\sum_{i=1}^{m} \sum_{j=1}^{n} y_{ij}/ \Sigma n_m)/\Delta x)_k \,| \;\leq\; \sum_{i=1}^{m} |\, \Delta((\sum_{j=1}^{n} \sum_{k=2}^{o} y_{ij}/ \Sigma n_m)/\Delta x)_k \,| \; / \; \# \text{ bins}.$$

If the condition that all derivatives are either positive or negative holds true, then

$$\sum_{k=2}^{x} |\, \Delta((\sum_{i=1}^{m} \sum_{j=1}^{n} y_{ij}/ \Sigma n_m)/\Delta x)_k \,| \;=\; 1/ \Sigma n_m * \sum_{k=2}^{x} \sum_{i=1}^{m} \sum_{j=1}^{n} |\, \Delta(y_{ij})_k/\Delta x \,|$$

which states that the absolute value of the average of a set of all positive or all negative derivatives is the same as the average of the absolute value of each derivative.

Substituting this relation into the average derivative equation and multiplying both sides by Σn_m, the equation of the average of the derivatives is the same as the equation of the individual derivatives. The algorithm for discretization of continuous chains is presented in figure 4-7.

---

1. Locate all continuous nodes in the continuous node chain
2. Identify a continuous root node of the chain
3. Sort the data in ascending numeric order on the root node
4. If duplicate values for the same set of input conditions exist for a value of the root chain node, average the duplicates
5. Find the derivative of each node in the chain with respect to the root

    $\Delta y_j / \Delta x = (y_{j+1} - y_{j-1}) / (x_{j+1} - x_{j-1})$ for j = 2 to m-1

    $\Delta y_j / \Delta x = (y_j - y_{j-1}) / (x_j - x_{j-1})$ for j = m
6. If all derivatives are positive or negative then

    $$\text{total change} = \sum_{k=2}^{o} \left| \Delta \left(\left(\sum_{i=1}^{m} \sum_{j=1}^{n} y_{ij} / \Sigma n_m \right) / \Delta x \right)_k \right|$$

    Else

    $$\text{total change} = \sum_{i=1}^{m} \sum_{j=1}^{n} \sum_{k=2}^{o} \left| \Delta y_{ijk} / \Delta x \right|$$
7. change per bin = total change / #bins
8. sum = 0 : i = 1
9. First cut point for each continuous node is the first set of values
10. sum = sum + $| \Delta y_i / \Delta x |$
11. If sum < change per bin, then i = i +1 and go to 10
12. percent = (change per bin – (sum – $\Delta y_i / \Delta x$)) / ($\Delta y_i / \Delta x$)
13. xcut = x_{i-1} + percent * ($x_i - x_{i-1}$)
14. For continuous node j, $ycut_j = y_{i-1j}$ + percent * ($y_{ij} - y_{i-1j}$)
15. sum = sum – change per bin
16. If #cut points < bins then go to 10, else go to 17
17. Last cut point is the highest value of the continuous node + 0.01 * highest value

| 18. If more continuous chains exist, go to 1, else stop |
| 19. If duplicate cut points exist, eliminate the duplicates |

Figure 4-7

Algorithm for Continuous Nodes with Continuous Parents/Children

4.1.2 Continuous Nodes with Discrete Parents or Children

If a network contains a continuous node that has no continuous children or parents, then the method described above must be modified. The principal concept of measuring the change over the entire data set and placing an equal amount of change in each bin remains the same. However, because there are no other continuous variables to work with the information contained in the discrete parents and children must be used. The derivative method is used to measure state transition changes in the discrete variables in relation to the continuous variable.

Figure 4-8 shows an example network with a continuous variable having no continuous parents or children. This example is created from a database collected from a California prison study. In the study, prisoners are assigned a classification score. The score is based on length of sentence, age, marital status and prior convictions. Prisoners are then assigned to a facility based on the score. Multiple offenders have higher scores and are usually sent to higher security facilities. The study then recorded if the prisoner committed any misconduct violations. The purpose of the study was to look at the score and the level of security of the facility as predictors of misconduct.

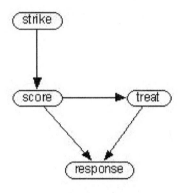

Figure 4-8

Prison Classification Network[5]

In this example, node "strike" is a discrete node with states "first", "second" and "third" that indicate whether a prisoner is a first, second or third time felon. Node "score" is a continuous node with values from 0 to 79 reflecting the prisoner classification score. Node "treat" is a discrete node with states "yes" and "no" indicating whether the prisoner was sent to a maximum security facility. Node "response" is a discrete node with states "yes" and "no" indicating whether this prisoner committed any misconduct violations while incarcerated.

To discretize the continuous variable, all parents and children are first identified. If a discrete node has states that are not numeric values, the states must be changed to numerical values. The numeric values are ordered such that variable states are assigned in the order in which they predominately occur in combination with the continuous variable. This is done by calculating the average value of the continuous values for each state of the discrete node. This is accomplished using the formula

$$avg_i = \sum_{j=1}^{n} state_j / n$$

i is the number of states of the discrete node

[5] Data obtained from the UCLA statistics website at http://www.stat.ucla.edu/projects/datasets/prison.txt

n is the total number of continuous values corresponding to discrete state i

state$_j$ is the value of the continuous variable at index j

avg$_i$ is the average value of the continuous values in discrete state i

The data is then sorted in ascending numeric order by sorting first on the continuous data and then on each parent or child respectively. Duplicate continuous data points are averaged as described in 4.2.1 to prevent an infinite slope. The derivative at each point is then calculated using the same formulae as described above. The cut points are found using the same method. Applying the derivative method to the example problem above for five bins provides the discretization shown in figure 4-9.

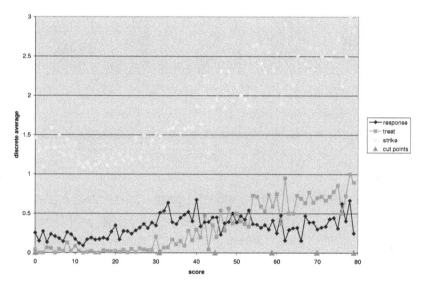

Figure 4-9

Five Bin Discretization for Prison Network

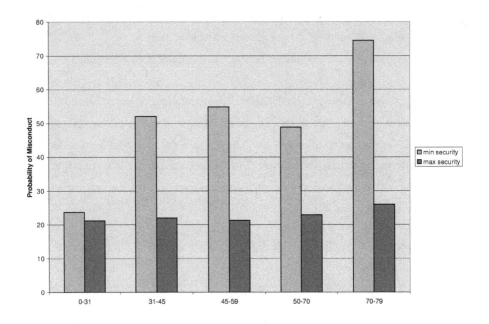

Figure 4-10

Prison Network Response

As can be seen from the figure, nodes "score" and "response" together are fairly accurate predictors of misconduct. With this discretization, the data shows that for the lesser security facilities, higher score does equate with higher misconduct. For maximum security facilities, the probability of misconduct is fairly constant over the range of scores. Misconduct for prisoners at the lower security facilities with low scores is about the same as those at maximum security facilities with high scores. Looking at the raw data of figure 4-9, one would conclude that misconduct is not strongly related to score as it seems to be fairly constant across the entire range.

The algorithm for discretization of continuous nodes with only discrete parents/children is presented in figure 4-11.

1. Locate the continuous node
2. Identify all parents and children
3. If discrete nodes are not numeric values, replace the states with numeric integers in ascending order of the average of the continuous values associated with the discrete state
4. Sort the data in ascending numeric order sorting first on the continuous node, and then on each parent and child
5. If duplicate values of the continuous node exist, combine multiple values of the other nodes by averaging
6. Find the derivative of each discrete node with respect to the continuous node
7. $\Delta y_j/\Delta x = (y_{j+1} - y_{j-1}) / (x_{j+1} - x_{j-1})$ for $j = 2$ to m-1

   $\Delta y_j/\Delta x = (y_j - y_{j-1}) / (x_j - x_{j-1})$ for $j = m$
8. total change $= \sum\limits_{i=1}^{n} \sum\limits_{j=2}^{m} | \Delta y_{ij}/\Delta x |$
9. change per bin = total change / #bins
10. First cut point = lowest value of continuous node
11. sum $= 0 : i = 1$
12. sum = sum $+ | \Delta y_i/\Delta x |$
13. If sum < change per bin, then $i = i +1$ and go to 12
14. percent = (change per bin $- ($sum $- \Delta y_i/\Delta x)) / (\Delta y_i/\Delta x)$
15. xcut $= x_{i-1} + $ percent $* (x_i - x_{i-1})$
16. For continuous node j, ycut$_j = y_{i-1j} + $ percent $* (y_{ij} - y_{i-1j})$
17. sum = sum – change per bin
18. If #cut points < bins then go to 12, else go to 19
19. Last cut point = highest value of continuous node + .01 * highest value
20. Stop
21. If Duplicate cut values exist, eliminate the duplicates

Figure 4-11

Algorithm for Continuous Nodes with Discrete Parents/Children

4.2 Limitations of the Derivative Method

The derivative method is very flexible and can accommodate linear, non-linear, continuous and discontinuous data sets. One limitation the method has is with data sets that are highly scattered. Because the method measures the total amount of change across a data set and then places an equal amount of change into each bin, poor results may occur if the algorithm mistakes a large difference between two closely spaced points for a rapid change in the slope. An example of a data set with high scatter is shown in the example of figure 4-12.

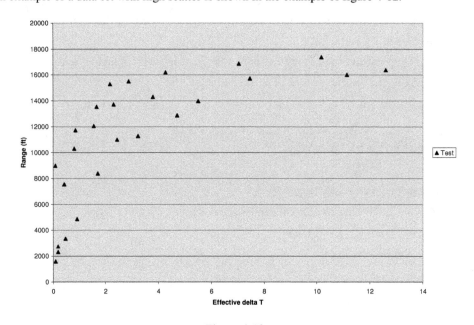

Figure 4-12

WF-360TL Infrared Receiver Performance Data

This limitation can be overcome by pre-processing the data prior to using it in the derivative method algorithm. The simplest method is to fit a curve through the data and then sample the function of the curve. This data can be used for input to create the bin discretization. After the network is constructed, the original data is used to learn the probabilities of the bin states of the node. An example of the discretization of a logarithmic curve fit through the data is shown in figure 4-13.

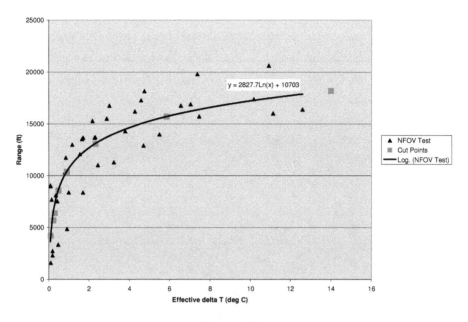

Figure 4-13

Cut Points for FLIR Data Set

Although fitting a curve to the data provides an acceptable solution to overcome this limitation, an alternative method that is used in this research is presented in chapter 5.

A second limitation is use with nodes whose data within a continuous node results in significantly different functions such as the input data shown in figures A-4 through A-8 of appendix A. This is not a limitation of the derivative method, but a limitation of Bayesian networks. A single continuous node can have only one set of ranges defining the discretization of that node. If data within that node is very different with respect to its slope, the derivative method will concentrate most of the cut points in areas such as a non-linearity while placing few cut points in linear areas of more gradual change. This results in very low resolution in the linear areas with corresponding poor performance. If this condition exists, an alternative approach is to group the data into sets where the response is similar and to create different models for different sets of data. This method is used for the amplifier model of appendix A which had both linear and non-linear inputs and outputs.

4.3 Substituting Probability Distributions into Continuous Node Tables

Another issue that must be addressed is the impact of the number of bins on the accuracy of the answer. As an example, the electrical circuit shown in figure B-1 of appendix B is tested for values of R = 1.8KΩ, C = 100μF and Time = 0. Figure 4-14 provides a normal distribution of the results of 26 tests.

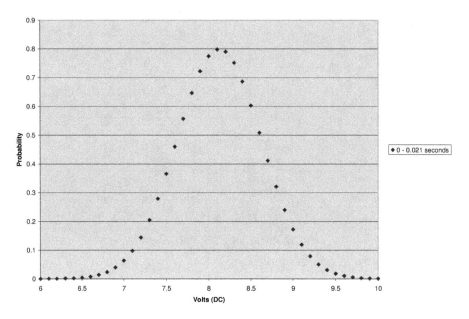

Figure 4-14

LRC Test Data Distribution

From this data the 95% confidence interval is between 7.1 and 9.2 volts. The average test data shown in figure 4-5 of eight different circuit configurations is used for Dirichlet learning of node probabilities in the network provided in figure 4-15. A ten bin discretization using the derivative method is used to create the network. The software adds one bin from negative infinity to the lower cut point and a second from the highest cut point to infinity. The addition of these bins is explained in chapter 6.

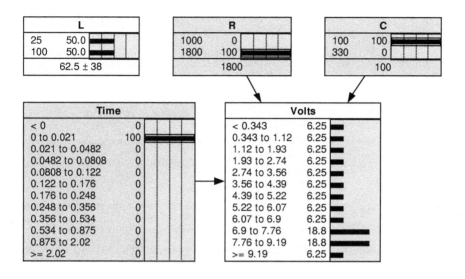

Figure 4-15

Dirichlet Learning with Uniform Prior

This results in a poorly performing network returning a mean of 5.37 for the node "Volts". As can be seen in figure 4-15, this number is not in either bin with the highest probability (6.9 - 7.76 or 7.76 - 9.19). This is because the learned network contains substantial probability on values less than 6.9 volts, although under a Gaussian distribution such values are extremely rare. This happens because the learning algorithm assumes the observations are multinomial and the prior distribution is uniform, resulting in a Dirichlet posterior distribution. The posterior bin probability is:

$$\text{bin probability} = N_{data} / N_{total} + 1$$

N_{data} is the number of data points that fall within a bin

N_{total} is the total number of data points.

The one is added to the denominator of all bins as a uniform prior assumes each bin has an equally likely probability prior to the introduction of the data set.

In figure 4-15, there are four values of node "Volts" that fall into the time range of 0.0 -
0.021 and have parents R = 1.8KΩ and C = 100μF. Two values fall within bin 6.9 - 7.76 and
two fall within bin 7.76 - 9.19. There are 12 bins that each start with one value in that bin.
Before learning begins, the probability distribution is uniform with a probability of 1 / 12 (0.083)
for each bin. After learning of the four data points, the probability is (1 + 2) / (12 + 4) or (0.188)
for both bins 6.9 - 7.76 and 7.76 - 9.19. The probability is 1 / (12 + 4) or (0.0625) for all other
bins. For a given number of data values as the number of bins is increased, the probability
within each bin that contains data decreases and the total probability in all bins which contain no
data increases.

There are three possible ways that were considered to correct this problem. The first
consideration is to conduct many more tests to obtain enough data points to reduce the
probabilities in the bins that contain no data values. However, extensive testing would be both
time consuming and expensive and would negate any advantages to using Bayesian networks for
modeling. A second option is to assign higher confidence to the data set by increasing the
number of times that each point is counted during the Bayesian learning process. If the network
of figure 4-15 relearns the probability tables from the learning case file data counting each case
one hundred times, the resulting probability for the node "Volts" is shown in figure 4-16.
Applying the same parent conditions now results in a mean value of 7.8 for node "Volts". This
results in improved performance with the mean value now within one of the bins with the highest
probability (7.76 – 9.19). Although improved, this option does not provide a completely
satisfactory answer. Based on the data, it is not likely that the probability of a value in bin 6.07 –
6.9 is the same as the probability of having a value in bin <0.343. The probability should be
much higher in the former as compared to the latter.

Learning performs so poorly because the learning algorithm does not make use of all the
information we have about the data. Specifically, the Dirichlet/Multinomial learning algorithm
applies to categorical data, while these data are continuous and known to be nearly Gaussian in
their distribution. The third method increases statistical power by including this information.
For this research, this is implemented by overwriting the CPT table of the continuous node with
a normal distribution obtained from the data set when the data is expected to be normally
distributed. To replace the distribution, the mean and standard deviation are calculated for all

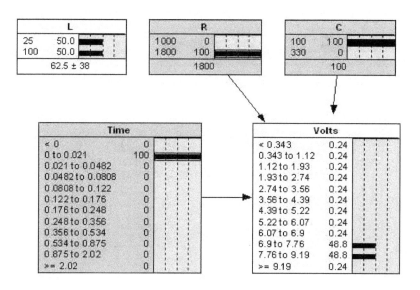

Figure 4-16

Dirichlet Learning with Non-Uniform Prior

data of a node for a single set of input states. In the circuit network example for R = 1.8KΩ, C = 100μF and continuous node "Time" bin 0.0 – 0.021 there are four values of "Volts" (two at time 0.0 with values of 8.60 and 8.61 and two at time 0.02 with values of 7.63 and 7.66) with mean 8.12 and standard deviation 0.50. The cumulative normal distribution is then calculated at each of the bin cut points. For the first bin (all values less than the first cut point) the probability is the cumulative normal distribution at the lowest point. For all but the last bin the probability is the cumulative normal distribution at the upper cut point minus the cumulative normal distribution at the lower cut point. For the final bin (all values greater than the highest cut point) the probability is one minus the cumulative normal probability at the highest cut point. This ensures that all probabilities for each row of the CPT table sum to one. An example of the node Volts using this method is shown in figure 4-17.

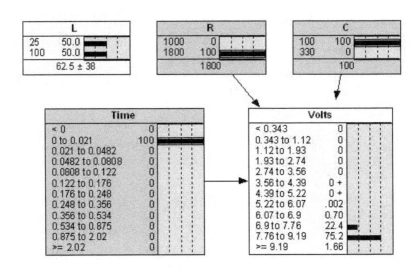

Figure 4-17

Normal Distribution Continuous Node

The value of node "Volts" at the reference conditions is 8.23 ± 0.66. This is very close to the mean of 8.12 of the data. As can be seen in figure 4-17, the lower bins now have zero probability and therefore no influence on the output. Use of this method allows discretization of continuous nodes using a higher number of bins to obtain higher model resolution without loss of accuracy. Either methods two or three will be used in this research.

4.4 Comparison of Discretization Methods

Recent research compared two search methods and two scoring methods for both time and accuracy [Xu, 2003]. The search methods were Hill Climbing (HC) [Pearl, 1985] and Markov Chain Monte Carlo (MCMC) [Gilks et al., 1996]. The scoring methods were Minimum Description Length (MDL) [Friedman and Goldszmidt, 1996] and Bayesian Dirichlet (BD) [Heckerman et al, 1995]. Combining the combinations of searching and scoring, this resulted in four different methods for the discretization process: Hill Climbing Minimum Description Length (hcmdl), Hill Climbing Bayesian Dirichlet (hcbd), Markov Chain Monte Carlo Minimum

53

Description Length (mcmdl) and Markov Chain Monte Carlo Bayesian Dirichlet (mcbd). The software to implement these four methods was written in MATLAB®.

The derivative method was compared to each of these four methods by means of two scoring metrics and the time it took the discretization algorithm to execute. The comparison was conducted on four data sets: the cycle amplifier data of appendix A, the LRC circuit data of appendix B, the aerodynamic wing lift data of appendix C and the California prison study data of section 4.1.2. The prison data contained only one continuous node, while all others contained more than one continuous node. Data were separated into mutually exclusive training data sets and test data sets. Three sets of training data and test data were constructed for each data set. The test set was determined by a random number generator which selected approximately 10% of the total data from each set. If data sets were ordered, the selection was made so as to evenly sample the test cases over the data set. This was done by randomly sampling equal numbers from equal time intervals on the amplifier and LRC data sets and sampling one angle-of-attack for each wing shape. Discretization was first conducted using the comparative methods which determine the number of bins using a user supplied maximum number and their respective scoring techniques for merging or splitting the data. The data were then discretized using the derivative method using the number of bins determined by the comparative method. All data runs were conducted on a 3.07 GHz computer with a Pentium 4 processor and 1.0 GB of RAM.

The comparison was conducted by first using the training data set to calculate the probabilistic relations between variables. The networks were then tested with the test data set and scored using two methods: Log Loss and Spherical Gain. The equations for these network scoring metrics are:

$$\text{Log Loss} = \text{MOAC} * (-\log(P_c))$$

$$\text{Spherical Payoff} = \text{MOAC} * \left(P_c / \left(\sum_{j=1}^{n} P_j^2\right)\right)$$

where P_c is the probability predicted for the correct state, P_j is the probability predicted for state j, n is the number of states and MOAC is the mean average over all cases[6]. The comparison attempted to use inputs of a maximum of 10 bins for each method and 25 iterations per

---

[6] Definitions from the Netica User's Manual.

discretization cycle. These numbers caused the computer to hang up and freeze for the wing and prison data sets. Because no error message was generated, it is suspected that the problem is linked to the large size of the network and file for these two data sets (wing has a variable with 5 parents and prison has 3917 cases in the data set). Successful discretizations were obtained by reducing the maximum number of bins to 5 and the maximum number of iterations to 10. The Log Loss comparison for the test cases is presented in figure 4-18 and the Spherical Payoff in figure 4-19.

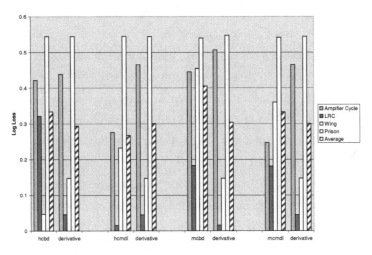

Figure 4-18

Log Loss Comparison of Discretization Methods

The results of this comparison confirm results from another study that indicate that different methods of discretization have different results on different types of data sets [Lui et al, 2002]. The results of this comparison show differences between the methods on individual data sets, but the average of the four cases for each method is nearly the same for both scoring metrics. Based on the test cases, the derivative method demonstrates approximately the same performance as compared to the four other methods.

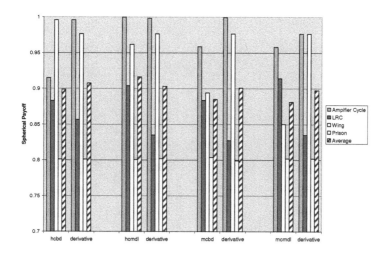

Figure 4-19

Spherical Payoff Comparison of Discretization Methods

The time required to execute the discretization algorithms was recorded for each method. The plot of time is presented in figure 4-20.

The results agree with a previous study that Markov Chain Monte Carlo methods take longer than Minimum Description Length [Xu, 2003]. In all cases the derivative method was faster than either of the other comparative methods. The magnitude of the difference can not be directly compared from figure 4-20 in that the comparative methods were run in MATLAB® while the derivative method was run in Microsoft Visual Basic®. Because the derivative method was running in a compiled program while the comparative methods were not, some of the speed difference must be attributed to the difference between the two programs. However, the magnitude of the difference is too great to be related only to the differences in code, and a conclusion can be made that the derivative method is faster than any of the other four methods. Further research is required to determine the exact magnitude of this difference.

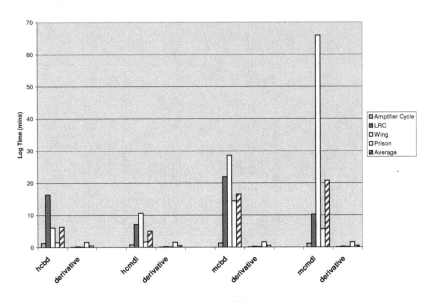

Figure 4-20

Time Comparison of Discretization Methods

4.5 Discussion

The derivative method of discretization was developed when a review of published
methods found none suitable for use in the discretization of continuous data for Bayesian
network engineering models. The methods reviewed either provided poor results during initial
research or were deemed too complex and inefficient to provide a viable alternative for
comparison to equation-based models. As such, the development of the derivative method is a
by-product of this research effort. Based on a limited comparison to a small sample of other
methods, this dissertation will not attempt to prove that this method is superior to any other
published method or is optimally implemented as described in this chapter. However, the results
obtained using this approach and described in chapter 7 demonstrate this method has strong
potential. There are a number of follow-on research projects that should be considered.

Although the method is computationally efficient, it may not produce the best
discretization of data sets that can be obtained. The method always results in the same number
of bins for every node in a continuous chain unless duplicate cut points exist. A better
discretization may exist by allowing different nodes to have different numbers of bins. A

potential follow-on research effort might look at using the derivative method using a large number of bins to create a starting discretization, and then using published merge methods to see whether a better discretization can be achieved with fewer bins. This might improve the time to converge to a solution for merge methods by providing a better starting point.

Different implementations of the derivative method rules should also be explored. The current implementation is to use equal-weighted averages for duplicate values of the same X variable. An alternate approach would be to explore the use of probability weighted averages. The method currently uses a manual input to determine the number of bins for data discretization. This was useful for the research in that in most cases, a fixed number of bins was desired for comparison of different modeling techniques. Finding a method to determine the number of bins for the derivative method is another area that should be explored. Automatic calculation of number of bins to achieve a given accuracy would be one selection criterion. Development of metrics which would minimize the number of bins for some criterion such as accuracy or resolution should also be considered.

It has been shown that informed methods of discretization provide better structural learning than uniformed methods [Monti, 1999]. The current implementation uses uniformed methods for initial discretization before structural learning, and then uses an informed method to discretize the data again once the structure has been determined. This may result in a good discretization of variables in a less than optimum structure. The software as implemented is computationally efficient and executes relatively quickly. An interesting area for future research would be to look at an iterative approach to structural learning where the informed discretization data is used to relearn the structure and then the data is discretized again on the new structure. The process would continue until it converges on one or more structures.

5. Building Models When Data are Incomplete

5.1 Adjusting Data Sets for Missing or Incomplete Data

When learning node probabilities from data sets, the quality of the final network will be influenced by the quantity and completeness of the data. Unfortunately, model builders may not have the luxury of ordering custom data sets and frequently must work with what is currently available. These data sets may contain cases in which the measurement was missed or was unavailable. Measurement of the data might have resulted in high concentrations of data in some areas with a scarcity of data in other areas of interest. Engineering trade studies frequently need data for configurations which have never been built or tested. To be useful as an engineering modeling tool, Bayesian networks must be able to model responses for conditions which have not been previously measured. In all of these situations, there is a need to generate predictions for missing or incomplete data.

There are a number of manual techniques that could be applied to augment the database. Adjustment by human judgment is one method that is specifically evaluated in this research. A second method is the use of computer predictions to augment the database. A computer algorithm creates a model of the available data and then uses the model to generate predictions. This method is referred to as "borrowing of strength" which refers to the use of available data to make predictions where data are sparse [Kreft, 1996]. This method is currently used in some predictive applications by the U.S. Census Bureau [Judson, 2002].

In order to make the predictions, a regression model must be created for the node or nodes for which values are needed conditional on their parents. Such a regression model would express a probability distribution of the node as a parametric function of its parents. The most common type of regression model consists of a deterministic function plus a normally distributed random noise term with zero expectation and standard deviation (σ). This can be expressed as

$$y = f(x,\theta) + N(0,\sigma^2)$$

y is the variable of interest

$f(x,\theta)$ is the equation conditional on the parent states x and θ

$N(0,\sigma^2)$ is the noise component normally distributed about mean zero

There are many methods that can be used to estimate the parameters of this regression model. One method would be to fit a curve through the data. This is an acceptable approach as long as all data for a node consists of continuous linear functions. Unfortunately, this is not the case for many engineering applications. Real data may be discontinuous such as the input data shown in figure A-9, or non-linear as shown in the input data of figure A-10 of appendix A. Attempts to fit curves to these data sets would result in a poor representation of the data set and consequent inaccurate predictions. In order to handle discontinuous or non-linear data, the approach taken in this research is to use a neural network to create a semi-parametric regression model of the available data. This model is then used to impute missing values for random variables.

5.2 Neural Networks

The principal advantage of a neural network is that it is capable of adaptive learning of very complex problems (Maren, 1990). Because this research effort is looking at alternatives to equation-based modeling, the ability to make predictions without creating equations to describe variable relations satisfies the research goals. These networks can predict additional values within the range of the training data set. Neural networks can also handle both non-linear and non-continuous functions. As implemented in the research, the neural network will be trained using the subset of the input data where all values are present for each case. Once trained, the network will be used to augment the input data prior to construction of a Bayesian network.

There are different types of neural network applications available for consideration. They fall into five basic categories: prediction, classification, data association, data conceptualization and data filtering. The primary use of a neural network in this research is for prediction. Types of predictive neural networks include the back-propagation, delta bar delta, extended delta bar delta, directed random search, higher order or functional link, and the self-organizing into back-propagation. Of these all use the feed forward back-propagation method of learning described below with some modifications associated with the specific type except for the directed random search network [Anderson and McNeil, 1992]. Feed-forward back-propagation networks (usually referred to as the back-propagation networks) are available in a variety of commercial applications. A back-propagation neural network was therefore selected for use in this research.

A typical back-propagation neural network is shown in figure 5-1.

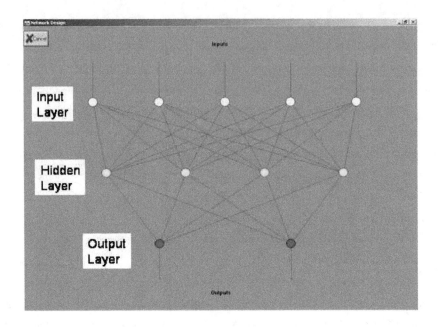

Figure 5-1

Example Neural Network

The network consists of an input layer, one or more hidden layers and an output layer. The input layer nodes feed the input values into the rest of the network. Connections between layers are bi-directional. Data values move from inputs through the hidden layers to the outputs during feed forward operation. During learning error corrections are propagated back through the network starting from the output nodes and running upward through all hidden nodes from the bottom to the first hidden layer.

All hidden and output nodes in the network have the structure shown in figure 5-2.

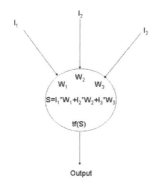

Figure 5-2

Neural Network Node

These nodes can have any number of inputs. During feed forward operation, the node first calculates the sum of all inputs times their weights. A transfer function is then applied to the sum. This function transforms the output into a number between zero and one (minus one and one in some software packages). There are several transfer functions that can be used. All functions have either a ramp, bell or modified S-shaped curve that runs asymptotically along the X-axis approaching either the maximum or minimum value [Maren et al, 1990]. The type of transfer function is manually selected during network construction while the weights for each input connection are calculated during the learning process described below. All inputs must also be scaled to values between zero and one. Outputs, which are all values between zero and one, must be scaled in the reverse direction from a decimal value to the actual value. Most software packages, including the one used for this research, perform the scaling task automatically so that it is transparent to the user.

The feed forward back-propagation method of learning iteratively adjusts the values of the weights for each hidden and output node during the learning process. In the first step, the input values are read into the input nodes and these values are fed forward through the hidden nodes until an answer is calculated at the output level as described above. The calculated output is then compared with the target output which is the value of that node in the training data given the input data. To correct the weights, the Generalized Delta Rule expressed in terms of a Delta function is used. This function is applied to every connection weight after each presentation of a training data set. The weights are then updated using

$$weights_{new} = weights_{old} + \alpha * Delta(weights_{old})$$

where α is the learning rate that controls how fast the weights are changed. The Delta function is proportional to the negative of the derivative of the error with respect to the connection weights [Maren et al, 1990]. The actual form of the Delta function is dependent on the transfer function chosen for the node. The greater the slopes of the error with respect to the weights, the greater the change that is applied to the weights. The error corrections are propagated backward through the network to update the weights in each node. This process continues until the weights in the network converge to values that minimize the error between the calculated values and the target values. The speed of convergence is affected by the learning rate alpha (α). Low values of alpha will cause the weights to be adjusted slowly resulting in slow convergence on final weights. A large value of alpha will result in faster convergence, but may cause the corrections to overshoot causing network learning to become unstable and the weights to diverge from the solution. When learning is complete the relations between variables are stored in the weights of the hidden and output layers of the network.

Unlike Bayesian networks, there are no software packages available that can learn a network structure from a data set. Construction of neural networks is primarily a manual operation. Like Bayesian networks, developing good neural networks can be as much an art as it is a science. The model builder must manually choose the number of layers, the number of nodes in each layer and the transfer function for each layer. These are all critical decisions which affect the quality of the final network. Increasing the number of layers and nodes in the hidden layers will improve the final fit of the network to the data. However, just as in curve fitting it is possible to "overfit" the neural network to the data set.

If a network is overfitted to the data, the network provides very accurate replication of the learning data set but poor performance predicting values between learning set points. In such cases, neural networks have poor generalization capability to any values outside the training data. This problem can be avoided by limiting the number of hidden layers and hidden nodes. Although a back-propagation network must have at least one hidden layer, 80% of all problems can be solved with a single layer [DeClaris and Roberts, 1997]. The most complex problem can be solved with three hidden layers [Anderson and McNeil, 1992]. Additional research has been

63

done to look at the total number of nodes in the hidden layers. For the functions evaluated in the research, the selection of two or fewer nodes resulted in underfitting of the data while selection of nine or greater nodes resulted in overfitting [Zhong and Cherkassky, 1999]. Many neural network software packages also allow designation of part of the input database as test cases. This data is then separated from the learning data set. The network is tested periodically during the back-propagation learning process with the test data used to check how well the network is learning the variable relationships.

Neural networks are usually fully connected such that every node in one layer has an input connection from all nodes in the layer above it and output connections to all nodes in the layer below it. If nodes have no relation, the network will automatically determine this during the back-propagation learning process and weighting values near zero will be calculated for connections that do not exist in the data relations. For this research, networks are constructed with the number of input nodes equal to the number of columns in the data set with no missing values. The number of output nodes is set to the number of columns with missing data values. The number of hidden layers is fixed at two. This number was determined by running tests on some of the data sets. The tests found that there was significantly better network performance using two hidden layers as compared to one, but no significant difference between two hidden layers and three. The number of nodes in each hidden layer is a step down value from the layers above except that the minimum number of nodes in the first hidden layer is two. This insures that each network has at least three total hidden nodes to prevent underfitting. Each network also has fewer than eight hidden nodes to prevent overfitting. All networks are hybrid networks with sigmoid transfer functions in the first hidden layer and the output nodes and Gaussian transfer functions in the second hidden layer as shown in figure 5-3. The transfer function type was determined by experimenting with the different functions and comparison of single transfer function networks with mixed hybrid networks. Although the combination sigmoid-Gaussian hybrid network performed the best, the difference was not highly significant. Other combinations of transfer functions provided networks that also performed well.

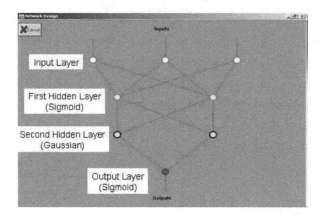

Figure 5-3

Research Neural Network Structure

5.3 Using Neural Networks in Bayesian Network Construction

The primary purpose of the neural network in this research is to impute missing values to estimate probability distributions for configurations in which there is no data. For continuous variables, this is a straightforward process with software sending the known values to the neural network as inputs and then retrieving the output variables which are then used to fill in missing entries in the database. Discrete variables require additional steps. All inputs and outputs of neural networks must be numeric values. Non-numeric discrete states are changed to integer values using the same approach as described section 4.1.2. If a predicted value is discrete, the output value from the neural network must be converted to one of the discrete states. If the discrete values are numeric, the values are rounded up or down to the nearest integer value. If the values are non-numeric, the neural network outputs are rounded up or down to the nearest integer value and then converted back to the non-numeric using the reverse of the process of 4.1.2.

A secondary use of the neural network is to eliminate the data scatter pre-processing requirement described in section 4.2. During learning, the neural network fits a numeric solution to the data. By using neural network predictions in place of the database values, the data set is effectively smoothed. An example of the method applied to the FLIR data set of figure 4-11 and then discretized using the derivative method is shown in figure 5-4. As can be seen, the cut points for figure 5-4 are very similar to the cut points in figure 4-13. For this research, if the neural network input option is selected, the output of the neural network is used for calculation of the bin dividing points using the derivative method described in chapter 4. If the data set is smooth, the neural network output will be very close to original data set with little impact on cut point selection. If the data set is scattered, the neural network smoothes the data insuring proper execution of the discretization algorithm. This eliminates the data scatter limitation of the derivative method allowing direct use of scattered data sets.

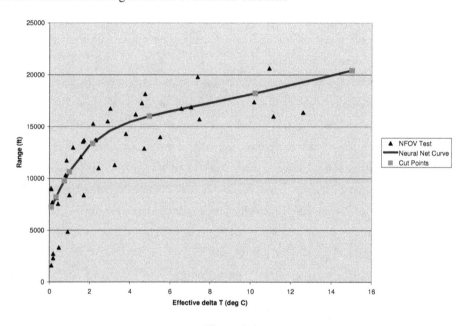

Figure 5-4

Neural Network Smoothing Example

5.4 Transforming Neural Network Outputs into Bayesian Network Probabilities

If there are missing values in the input data set, those values will be calculated by the neural network resulting in a new data set containing a mixture of the input data set and neural network predictions. This data can be used to provide a larger number of data values for each bin to reduce the influence the number of bins has on output probabilities if Dirichlet learning is used. It also can be used to ensure that even if the original data set has gaps that would result in

66

bins that have no data values, the final learning set will have the gaps filled in by the neural network predictions. Two methods are considered for handling the mixture of learning data with neural network predictions as described below.

The first method is to create a very large data set by using an input data set with a large number of missing data values. This data set is then used as the training data for Dirichlet learning of the probabilities. Although viable, this method has significant shortcomings. Calculation of a very large data set incurs significant computational time penalties and is therefore not efficient. The final probabilities will still have residual values in bins far from actual data values in the learning set as described in section 4.5. Finally, the probabilities do not take into account the additional uncertainties introduced into the output data set by the introduction of neural network predictions.

A second and better solution is to calculate the uncertainty between the neural network predictions and the training data and add it to the uncertainty of the data values within each bin. This method allows direct calculation of probabilities without generating overly large data sets. It also includes the uncertainty due to the errors introduced by the neural network predictions. Using the normal distribution substitution method described in section 4.4, the standard deviations are calculated for the data points that fall in each bin of a designated continuous leaf node. The variance of the data within each bin is then calculated by squaring the standard deviation. The uncertainty of the neural network data can be calculated as a variance using the formula

$$\sigma_{nn} = (\text{Learning Data} - \text{Prediction})^2$$
σ_{nn} is the variance of the neural network data
Learning data is the value of a point in the input data set
Prediction is the neural network output for the same set
of inputs as the Learning Data

This results in a variance for each point in the learning data set. The model builder is then faced with a choice of which variance or combination of variances to use for the uncertainty due to the neural network prediction. For this research, three options are considered for determination of this variance: nearest value, average and maximum.

The nearest value option searches the input data set and finds the variance associated with the set of input values that is closest to the input values used with the neural network prediction.

This option is most suitable for data that is not scattered and where it is reasonable to assume that the error in the neural net output will be similar under like conditions. An example is shown in figure 5-5.

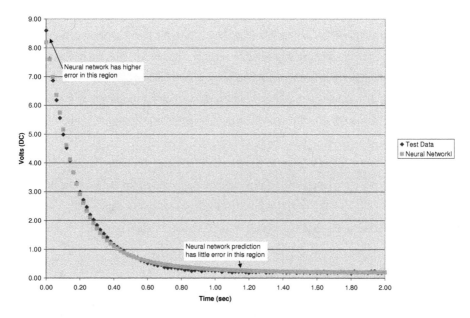

Figure 5-5

Neural Network Prediction Accuracy

The nearest value option provides the most accurate prediction of error under the specified assumptions. However, this option requires significant computer run time to execute if the number of missing values is large and/or the database of complete data contains a large number of values.

The average variance option is best suited for use if the data set is scattered. In this case there may be measurement errors in the input data set. The average variance provides a reasonable variance estimate across the entire data set under such conditions and prevents local extremes that would occur if the nearest value method is used. The average value option would be appropriate for input data such as that shown in figure 5-4. The maximum variance is the highest variance found between the learning data set and the neural network prediction across the entire data set. This method is used with data sets which are not scattered. This provides the

68

most conservative error estimate of the three methods. Both the average and maximum options are more computationally efficient as compared to the nearest value option.

Once a variance is selected, a new normal distribution is calculated that incorporates sources of uncertainty from both the data and the neural network. If all values that fall within a single bin for a unique set of inputs are data values from the learning set (i.e. no neural network predictions), then the procedure described in section 4.3 is used. However, if any values within the bin range are generated by the neural network, a new standard deviation that includes both the variation of data values within each bin plus the variation introduced by the neural network predictions is calculated. The formula for calculating the total standard deviation is

$$\text{std dev}_{total} = ((\text{std dev}_{data})^{\wedge}2 + \sigma_{nn})^{\wedge}0.5$$

std dev$_{total}$ is the standard deviation for both sources of uncertainty

std dev$_{data}$ is the standard deviation of the data for the inputs

σ_{nn} is the variance of the neural network data

The normal distribution is then calculated using the total standard deviation and the probabilities replaced in the node probability tables as described in section 4.3.

5.5 Discussion

Integration of neural networks with Bayesian networks combines the strengths of both artificial intelligence agents while eliminating some of the individual limitations of each. The neural network allows interpolation of data values not contained within the learning data set. It provides smoothing of scattered data sets effectively mitigating one of the limitations of the derivative method described in section 4.2. In combination with the discretization methods described in chapter 4, this allows the creation of engineering models that can predict outputs of any set of input values bounded by the learning data set. Using a Bayesian network as the final product allows reasoning with incomplete inputs. Predicted outputs are qualified as probabilistic distributions providing a range of values over which the output may lie. This research supports a general trend in which the areas of neural networks, statistics, generative models and Bayesian inference are coalescing into a single field of Soft or Natural Computing focused at drawing conclusions from incomplete, noisy data [Smith, 2001].

Implementation in this research uses a single, manually constructed neural network for prediction. This represents one of the few, non-automated tasks in the Bayesian network construction process. Additional research should be conducted to explore computer generation of the neural network structure. While development of an algorithm to select the input and output nodes would be straight forward, selection of the number of hidden layers and number of nodes in each of those layers is the challenge. The algorithm must choose a network structure that provides enough layers and nodes to adequately fit the network weights to the training data without making the network any more complex than is necessary. One method that has been developed to address this issue is Vapnik-Chervonenkis (VC) theory used for estimating data dependencies from finite data samples. This provides a framework for choosing one model from a set of possible models using Structural Risk Minimization (SRM). SRM scores and orders the model set according to their complexity. Model selection is then accomplished by choosing a minumal analytic upper VC bound of the prediction risk [Zhong and Cherkassky, 2000], [Cherkassky and Mulier, 1998]. Use of multiple regression models is another area that has been explored in recent research. Use of a single neural network assumes that all data was generated by a single, unknown model. Better solutions may exist by modeling the data with two or more regression models [Cherkassky and Yunqian, 2002a] [Cherkassky and Yunqian, 2002b].

6. Research Software Implementation

To conduct the proposed research, a software package capable of creating Bayesian network models from data sets is required for comparison to manually constructed networks. A search of existing applications found no software package suitable for creating engineering models from data sets containing mixtures of discrete and continuous variables. The primary deficiency was the absence in available packages of methods for intelligent, simultaneous discretization of multiple continuous variables. This led to the development of the derivative method of discretization which is implemented in the research software package described below. The software package, BN Builder, integrates new code to implement the procedures described in chapters 4 and 5 with four third-party software packages providing the rest of the functionality required to construct a Bayesian network from a data set. Integrating existing software to the maximum extent possible minimized the amount of new code development required to perform the research tasks with respect to computer constructed models. The software architecture and data flow are presented in figure 6-1.

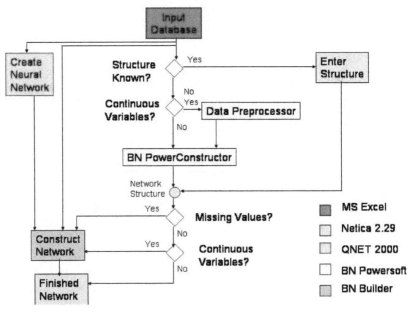

Figure 6-1

Integrated Software Architecture

71

6.1 Supporting Software Applications

Microsoft® Excel is used as the database input file format. This is a natural choice because many data sets are stored in this format. Furthermore, because Excel can read databases stored in many formats, there is minimal need for manual transformation. The Excel workbook also provides an output record of processes performed during execution of the BN Builder program. During the build process, extra worksheets are added to the Excel workbook and additional data is generated and written to these worksheets. This not only assisted in debugging during software development, but allows anyone constructing a model to see the details of the numerous intermediate operations carried out during construction. Excel also provides built-in functions for many of the operations required during the build process such as sorting, calculation of various statistics and generation of distributions.

Bayesian network structure learning was performed by BN PowerConstructor, one of the modules in the BN PowerSoft collection by Jie Cheng of the University of Alberta, Canada. Because BN PowerConstructor works only for discrete variables, BN PowerSoft provides a Data Preprocessor program which discretizes continuous variables using either the equal frequency or equal width method of unsupervised discretization. The BN PowerConstructor program constructs belief networks by using conditional independence (CI) tests. In general, it requires CI tests to the complexity of $O(N^4)$; when the attribute ordering is known, the complexity is $O(N^2)$ where N is the number of attributes (fields) [Cheng et al., 1997]. The resulting structure can be edited within the PowerConstructor program and saved in most common formats used by commercial Bayesian network software packages.

Output networks created using BN PowerConstructor are not suitable for use in this research. The first problem is that the nodes are discretized using uninformed methods resulting in poor performance as previously discussed in section 3.3. A second problem is that the output networks contain all discrete nodes with non-numeric or string state names. All numeric values are translated into strings by placing an "x" in front of a truncated portion of the number. Discrete numbers are truncated to the three right digits, while continuous numbers are truncated to the left three digits. This results in a discrete numeric node with states of 1000 and 1800 in the data set having states of x000 and x800 in the output network. This not only prevents the states from being set by entering numeric values, it is confusing even when manually entering and retrieving the state values. This also precludes the integration of these networks with other

Bayesian networks that use numeric values. Because of these limitations, networks created using BN PowerConstructor are only used to provide the relationships between variables of the network.

Netica, by Norsys Software Corp., serves two functions in the architecture: input of structural relations and output of the finished network. Netica version 2.29, an experimental beta version not yet available to the general public, is used for this research. This version is used because it contains the command set for handling of continuous variables in the Visual Basic programming environment; commands not available in the current commercial version of the software. If the relationship between variables is known, the structure is entered into Netica using the graphical user interface. The user enters the nodes, names the nodes with the variable names that match the column labels in the input database and draws the arcs showing the relationship between variables. The input Netica file, whether created by BN PowerConstructor or manually entered by the user, only provides variable relationship information. All other information is discarded by the BN Builder program. Therefore, no variable type, state names, ranges, etc. needs to be entered during manual construction.

The output Bayesian network is reconstructed in Netica format by the BN Builder software package. Netica version 2.29 contains a COM interface with the Netica engine. This provides a software command set that allows any operation that is supported in the graphical user environment to be invoked within the Visual Basic program environment. The BN Builder software makes use of these commands to completely reconstruct the network using the algorithms described in chapters 4 and 5. The output is a complete and ready to use Bayesian network file.

Qnet 2000 by Vista Services, Inc. is a software program for constructing, training and recalling predictions using neural networks. This package comes with a graphical user interface for constructing and training back-propagation neural networks. The software also comes with a Dynamic Link Library (DLL) file that provides software commands that can be used to send and receive data from the neural networks within the Visual Basic programming environment. Neural networks are created manually as described in section 5.2. The neural network is an optional input to the software build process. During program execution, BN Builder sends the input values to the neural network whenever predictions are required. The neural network generates the output variable predictions for the input data values which are then returned to the

73

main program. These values are stored in the Excel workbook for use as needed in the build program. Neural network predictions are tagged with a blue background color in the spreadsheet cell so that the program can identify which values of the database are inputs and which values are neural network predictions. This information is needed for calculation of the node probabilities if the normal distribution option is selected as described in section 5.4.

6.2 BN Builder Research Software

Microsoft® Visual Basic 6.0 was selected as the programming language for integration and custom component software because it can be compiled into an executable program for rapid execution, is compatible with integrating the supporting applications mentioned above and is the language most familiar to the developer. The application was constructed using a spiral development process and modularization of the functionality. Extensive documentation is included in the form of comments within the software code. The modules are stand alone functions such that there are no subroutine calls within one subroutine invoking another subroutine. This allows the direct addition or substitution of other functions and subroutines, even those written in other languages, so long as they conform to the Visual Basic calling convention. Although the program was specifically designed for this research, the architecture was designed to accommodate easy insertion of additional functionality or modification of the program for future research.

6.2.1 Graphical User Interface

The user interface screen provides the user selectable options for Bayesian network construction. The input screen also provides outputs that are useful for monitoring of the construction process. The user interface screen is presented in figure 6-2.

The functionally of each control on the user interface page is described below:

- "Build Status" window - provides feedback to the user on which of the operations described below is currently being executed by the program. The window is updated each time one module finishes and the next one begins execution.

- "Maximum Number of Discrete States" - up-down select button allows selection of the threshold for determining if a node is discrete or continuous. If a node has a number of states less than or equal to this value, the node will be treated as a discrete node in the final network. If the number of states is greater, then the node will be treated as a continuous node.

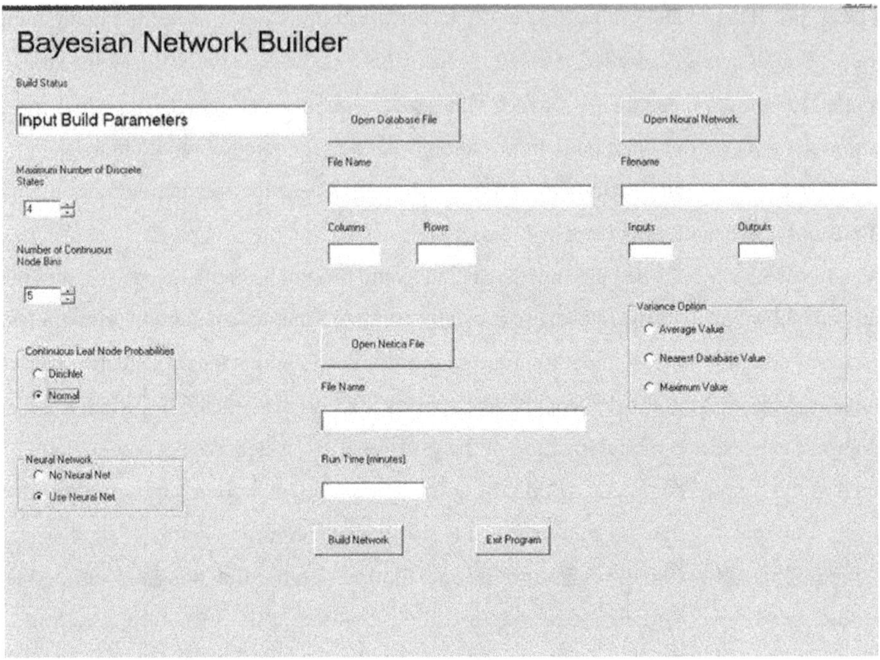

Bayesian Network Builder

Build Status

| Input Build Parameters | Open Database File | Open Neural Network |

Maximum Number of Discrete States

`4`

File Name

Filename

Number of Continuous Node Bins

`5`

Columns   Rows

Inputs   Outputs

Continuous Leaf Node Probabilities
- ○ Dirichlet
- ● Normal

Open Netica File

File Name

Variance Option
- ○ Average Value
- ○ Nearest Database Value
- ○ Maximum Value

Neural Network
- ○ No Neural Net
- ● Use Neural Net

Run Time (minutes)

| Build Network | Exit Program |

Figure 6-2

Graphical User Interface

- "Number of Continuous Node Bins" - up-down select button selects the number of bins
  into which the data for continuous variables is discretized when the derivative algorithm
  is executed. After the determination of the bin cut points, the software will always adds a
  bin from negative infinity to the lowest cut point and from the highest cut point to
  positive infinity. Because the derivative method places cut points based on change, it
  does not guarantee that the maximum and minimum values from the data set are captured
  when determining the lowest and highest cut points. The addition of these two ranges
  ensures that no errors occur if data values outside the maximum and minimum limits of
  the bins are present during learning of the probability tables.
- "Continuous leaf node probabilities" – determines the probability distribution in
  continuous leaf nodes. If the "Dirichlet" option is chosen, then Dirichlet learning with a
  uniform prior will occur. If the "Normal" option is selected, then the final probability

tables for continuous leaf nodes will be overwritten with a normal distribution with parameters estimated from the data as described in section 4.3.

- "Neural Net" – determines if a neural network will be used to generate missing data values and smooth the input data set. If the "Use Neural Network" option is selected, then the "Open Neural Network" option becomes visible on the right side of the screen. This option is used to specify that the user has constructed a neural network and wants it to be used during network construction. If the "No Neural Net" option is selected, the input data must not contain any blank cells. The open option will not be visible and the input data set values will be used for discretization.

- "Open Database File" - used to open the Excel spreadsheet containing the data set for the model. The data must on "Sheet1" of the workbook. The names of the nodes must be in the first row of each column and there must be no completely blank rows or columns within the data set. During execution, the number of rows and columns found in the input database is displayed.

- "Open Netica Network" - used to open the file containing the graph representing the dependency structure among the variables. This file must be in a format compatible with the Netica software application. The nodes in the network must exactly match the column labels in the database input file. The one exception is that when BN PowerConstructor discretizes variables, it adds a "_d" to the end of the variable name. The BN Builder software recognizes this tag and will automatically change its name back to the original.

- "Open Neural Network" - used to open a neural network. This option is only visable if the "Use Neural Network" option is selected. The network must be in a format that is readable by the Qnet 2000 software program. The input and output node names must match the column labels in the input database. The input data columns must be ordered so that the input nodes come first with the output nodes at the end. If the network is successfully opened, the number of input and output nodes is displayed.

- "Build Network" - executes the program. This option is selected after all the above inputs have been entered. When the software has completed the network construction, a "Done" message appears in the "Build Status" window.

- "Exit" - closes the BN Builder program and all other support applications. The user is provided the option to save the Netica output file and the Excel workbook including all the extra worksheets added during the build process as each of these applications closes.

- "Run Time" - displays the number of minutes from the time the user interface is first displayed until the final network construction is completed.

6.2.2 Software Execution

The software code executes a series of functions as shown in figure 6-3.

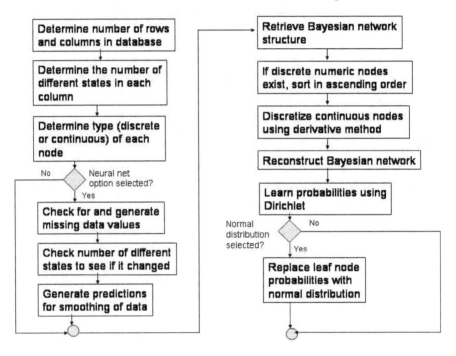

Figure 6-3

Software Execution Sequence

All modules in figure 6-3 are individually called functions in the software package with the exception of the last three which are all part of the network reconstruction operation. Program execution begins when the "Build Network" option is selected from the user interface. The program first checks the input Excel database file and counts how many rows and columns are in the database. It next checks to see how many unique states are contained in each column. This information is then used to determine whether each node is discrete or continuous in the final network. This determination is made based on the user set threshold for defining the maximum states allowed for a discrete node.

Once all node types are identified, the program will check to see if the neural network option is selected. If it is selected, the program looks for any cells with missing data. If empty cells are found, the software sends the available data in each row as inputs and retrieves the outputs which are then placed in the empty cell(s). Non-numeric discrete states must first be transformed into integers using the process described in section 4.2.2. Outputs for discrete variables are transformed back into the discrete state values for numeric states or the descriptive states for non-numeric values using the reverse of the same procedure. After empty cells in the database are filled in with predictions, the database is rechecked to see if the number of unique states has changed. If the neural network option is selected, the program adds a new worksheet to the Excel workbook and will create a new data set by sending the input values to the neural network values and retrieving network predictions for the outputs. This new data set is then used for discretization and insures that the derivative method is not influenced by scattered data sets as described in section 5.3.

The next operation is to open the Netica input file and retrieve the network structure. The program creates a matrix of the arcs identifying the variables and direction of the arcs. All other data in the network is discarded. The variable relations are used to set the arcs when the network is reconstructed, and is also used for the derivative method discretization process to identify the continuous node chains. If discrete numeric states exist, the data is sorted in numeric order. This operation insures that when the final network is constructed that the discrete numeric state nodes will have node states that go from the lowest number to the highest number. The software then calculates the ranges of all continuous variables by discretizing the data in the database. If the neural network option was not selected, the program discretizes the data set in the input data set. If the neural network option is selected, the program uses the neural network predictions for the outputs of the neural network in place of the data set values. The program then reconstructs the Bayesian network.

To reconstruct the network, the program first opens the Netica software application. It then inputs the nodes in the order they appear in the spreadsheet placing them in offsetting rows on the screen. In addition to the node name, each node is declared discrete or continuous according to the earlier determination of node type. Arcs are then added between the nodes based on the structure retrieved from the input network. If a node is discrete, the states of the node are set based on the states found in the input data set. If a node is continuous, the ranges of the bins are set based on the discretization results. The program then uses Dirichlet learning with a uniform prior to calculate the probabilities in each node from all the cases in the data set which includes the input data plus all predictions generated by the neural network. The software also saves the data in a file having the Netica case file format. This file can then be used to relearn the node probability tables as described in section 4.4 when a non-uniform prior is desired. If the "Dirichlet" option is selected, the network is finished and the program will show a "Done" message in the status window. However, if the "Normal Distribution" option is selected, the program will recalculate the probabilities of all continuous leaf nodes.

The normal distribution option is implemented to change the probability distribution of continuous leaf nodes. Only leaf nodes are implemented at this time due to the nature of the research. Because the research is primarily comparing engineering models, the root nodes are usually the inputs to the system and the leaf nodes are the outputs. The data sets are expected to have Gaussian distributions of the output variables. The software is implemented to provide the option of replacing the distribution of continuous leaf node probability tables with a normal distribution estimated from the data. This option provides a more accurate probability distribution in the node as explained in section 4.3. Additionally, if neural network predictions are used, uncertainties introduced by inclusion of neural network predictions are also included. This implementation is expected to maximize the predictive accuracy of the output Bayesian networks in both the central tendency and the spread of output solutions.

6.3 Extend Bayesian Network Integration Blocks

Another aspect of this research is to explore the integration of equation-based models with Bayesian network models. In this research, equation-based models and simulations are constructed and run using the Extend modeling and simulation software package. This package comes with a variety of capabilities including animation for visualization of models or simulations. It also comes with extensive libraries of modeling blocks that can be linked together to create a model. Neither this software package nor any other M&S software package investigated contained the capability to integrate a Bayesian network into a model. This required the development of a custom library of function blocks to enable the Bayesian network integration within the Extend modeling environment.

Each block in the Extend libraries is a visual representation of computer code written in a language called ModL. Users who are proficient in code writing can create their own blocks using this language. The Netica Bayesian network software package, chosen for this research, comes in two versions: a graphical user package previously described in section 6.1 and a set of C language Application Programming Interface (API) functions. Unfortunately, it is not possible to call the Netica API functions directly from the Extend modeling environment. This is due to the ModL language using a Pascal format for strings while the Netica functions use a C format. Although numbers can be passed, no alphabetic characters are compatible. This problem was overcome by developing a number-character translator written in C++. To access a Netica API function, the Extend ModL code first translates all character strings to their ASCII code values and creates a numeric vector of the numbers. The code then calls a C++ function passing the vector. The translator converts the numbers back to characters, assembles them in strings and then invokes a Netica API function. Only numbers are returned by the Bayesian network to the Extend modeling environment. If a return value requires a string answer such as the most likely node state for a discrete node, the answers is returned as the state index number and then translated to the state name in ModL.

In order to accomplish this integration, a custom set of blocks was developed to invoke one or more of the Netica API functions. These blocks provide most of the functionality of the graphical user software package. The blocks are color coded to allow for easy identification of the type of function it performs as described below. Red blocks are used to open and compile a network.

Start Network

Start Network – Initializes the API with the user license, opens a network and complies the network.

Open Network

Open Network - Initializes the API with the user license and opens a network. This block is used for networks with no entries in the probability tables.

Compile BN – Compiles the network.

Light blue blocks are used to send data from an equation model to the Bayesian network.

Set Node State

 Set Node State – Sets a discrete node to a specific state by state index number.

Set Utility Value

 Set Utility Value – Sets the utility value in a utility node.

Set Binomial Finding

 Enter Binomial Finding – Enters a finding into a discrete node with two states.

Set Node Value

 Set Node Value – Set the state of a discrete or continuous node with a numeric value.

Enter Finding

 Enter Finding – Set a discrete node state by state name.

Dark blue nodes are used to update the network after states have been entered.

Update Node
Probabilities

 Update Node Probabilities – Updates the node probability tables based on the current states of the nodes in the network.

Retract Findings

 Retract Findings – Resets all node findings that have been entered.

Green nodes retrieve data from the network for use by the equation model.

Get Belief

 Get Belief – Returns the probability that a discrete node is in a specific state.

Get Mean – Returns the mean value of a continuous node.

 Most Likely Value – Returns the mode of a continuous node.

 Get Node Statistics – Returns the mean and standard deviation of a continuous node distribution.

 Get node value – Returns the numeric value of a discrete node.

 Get Decision – Returns a decision from a decision node.

 Most Likely State – Returns the numeric state index of the most likely state for a discrete node.

Yellow blocks are used to save or close a network.

 Save Network – Saves the network under a new name.

 Close BN – Closes the network and frees the memory used by the network.

6.4 Discussion and Future Research

There are numerous improvements and future research studies that could be done with the BN Builder software. Some potential areas include algorithms to determine the number of bins for discretization, integration of the BN PowerConstructor engine into BN Builder, interleaved structural learning and discretization, and computer generated neural networks. The current software uses a manual entry to determine the number of bins for continuous variable discretization. This was useful for creating networks with a specific number of bins for comparison to networks created using other methods. However, if network storage size is an issue, it would be useful to develop metrics which balance the number of bins with accuracy, resolution or some other network parameter.

BN PowerConstructor is currently implemented as a separate, stand alone program. The program is executed once at the beginning of the process as shown in figure 6-1 only if the variable relationships are not known. BN PowerConstructor is implemented with an Application Programming Interface (API). This would allow direct integration of the structural learning functions within the BN Builder program. This could reduce the time required to execute the construction process. It would also allow interleaving of structural learning with discretization. This could potentially improve the final structures by alternately discretizing the continuous data based on the structure and then relearning the structure based on the discretization until the network converges.

If the neural network option is used, the network must be constructed and trained before it can be used as an input into the Bayesian network construction. Construction of the neural network by computer would further reduce the human time required during computer generation of Bayesian networks. Algorithms that produce multiple networks and then choose the best representation of the training data or algorithms that pick a best method based on the number of variables and type data are possible areas for future research.

The neural network integration is currently implemented to generate missing data for the output(s) of the system being modeled. This requires a single neural network to make the predictions. Some problems might have input data sets missing data values in all of the columns. In order to predict missing values under these circumstances, multiple neural networks could be used. Each neural network would have the variables with data values as inputs and the missing variable(s) as outputs. The software could be implemented to choose the appropriate neural network for each instance of a missing data entry.

The Bayesian network integration library and translator software was developed to demonstrate the ability to integrate Bayesian networks and equation-based models. Although successful, the current implementation is far from optimum due to the incompatible data formats of the two software packages. Execution is slow due to the multiple translations required to pass data back and forth between the two software packages. A different modeling and simulation package has been identified that uses a C format for its variables. Any follow on research will investigate a more direct integration with the Netica API functions.

7. Comparison of Modeling Methods

The primary purpose of this research is to compare the time of construction and accuracy of equation-based models to Bayesian network models. The research uses equation-based model results as the baseline for comparison to three different methods of Bayesian network construction: manually created networks with probabilities assessed by human judgment, manually constructed networks with formulae used to create probability tables, and computer-constructed networks using the software described in chapter 6 learning both structure and probabilities from data. The three types of models are compared to the equation-based baseline in both accuracy and time to construct resulting in the six hypotheses listed below.

Hypothesis #1A
Null hypothesis: Human judgment Bayesian networks have the same average percent error as equation-based models.
Alternate hypothesis: Human judgment Bayesian networks do not have the same average percent error as equation-based models.
Hypothesis #2A
Null hypothesis: Human judgment Bayesian networks require less than or equal time to construct compared with equation-based models.
Alternate hypothesis: Human judgment Bayesian networks require greater time to construct compared with equation-based models.
Hypothesis #1B
Null hypothesis: Formulae Bayesian networks have the same average percent error as equation-based models.
Alternate hypothesis: Formulae Bayesian networks do not have the same average percent error as equation-based models.

Hypothesis #2B
Null hypothesis: Formulae Bayesian networks require less than or equal time to construct compared with equation-based models.   Alternate hypothesis: Formulae Bayesian networks require greater time to construct compared with equation-based models.
Hypothesis #1C
Null hypothesis: Computer-generated Bayesian networks have the same average percent error as equation-based models.   Alternate hypothesis: Computer-generated Bayesian networks do not have the same average percent error as equation-based models.
Hypothesis #2C
Null hypothesis: Computer-generated Bayesian networks require less than or equal time to construct compared with equation-based models.   Alternate hypothesis: Computer-generated Bayesian networks require greater time to construct compared with equation-based models.

All formula models were created using formulae from published textbooks on the subjects being modeled, or derived from these published formulae. Human judgment was tested using a group of nine volunteer test subjects holding a Master's Degree or Ph.D. in math, science, engineering or engineering management. Subjects were provided background information and test results for similar items they were asked to predict as described in the appendices for each model. The computer-generated models used the same background data provided to the human judgment subjects to create an output model. For both the human judgment and computer-generated Bayesian network model accuracy tests, none of the data used for comparison to the test data was contained in either the test data shown to the volunteers or the input database used by the computer programs. Because the background data came from a variety of sources, it was impossible to measure the time to collect the test data. It is assumed that the human judgment and computer-generated models were using data that is already available. All computer-generated models were created using a 3.07 GHz Pentium 4 processor with 1.0 GB of RAM.

7.1 Comparison of Equation-Based Models with Human Judgment Bayesian Networks

The comparison of accuracy is made by comparing errors in the models of appendices A, B, and C. The three models are used to generate seven cases for testing the first hypothesis. A summary of model errors is presented in figure 7-1.

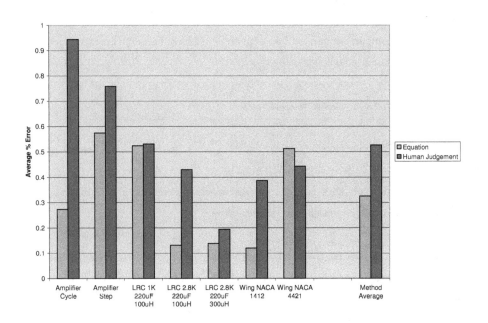

Figure 7-1

Human Judgment Model Error Comparison

Hypothesis 1A is tested with the data of figure 7-1 resulting in a failure to reject the null hypothesis at the 95% level of confidence. The error associated with the human judgment Bayesian network models is higher in six of the seven cases as compared to the equation-based models. Although the average error for the Bayesian networks is higher, the magnitude of the error is not great enough to establish a statistical difference between the two at 95% confidence.

Three cases are of particular interest in figure 7-1. The NACA 4421 wing is the only case in which the human judgment is better than the equation model. In this particular case, the models predicted the coefficient of lift of a two dimensional wing section. Humans were able to see and predict that some flow separation would occur at higher angles of attacks resulting in a loss of lift while the equation model could not as shown in figure C-12 of Appendix C. The other two cases of interest are the LRC 2.8KΩ 220μF 100μH and the LRC 2.8KΩ 220μF 300μH electrical circuits. For the values of the components selected, the value of the inductor has no observable effect on the output voltage as can be seen in figure B-4 of appendix B. The inductor causes a delay in the rise of the voltage when it is first applied. However, this delay happens so quickly it can not be observed within the maximum 0.02 second sampling rate of the test equipment for commercially available inductors. The prediction for the two circuits should be the same and this is correctly predicted by the equation-based model. By contrast, the human judgment prediction is significantly different as is shown in figures B-6 and B-7 of appendix B and is reflected in different accuracies in figure 7-1. This demonstrates biases which are present in human judgment predictions.

The times of construction for the three models that are used to create the seven cases are first presented in figure 7-2.

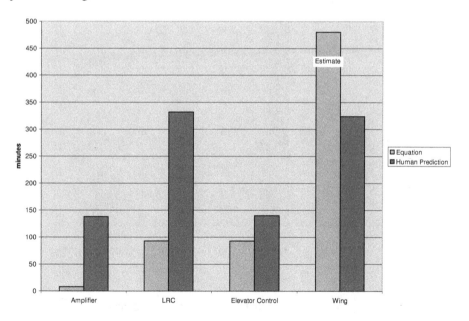

Figure 7-2

Model Construction Time Comparison

As can be seen from figure 7-2, time of construction is greater in three of four cases for human judgment models as compared to the equation models. The only case where the human prediction model construction time is less is the wing model. The wing model is shown only for comparison and is not used for hypothesis testing as it was obtained from an outside source that could only provide an estimate of construction time.

In addition to the human judgment models shown in figure 7-2, an elevator control model and a thermostat model described in appendices D and J were also constructed for use in chapter 8. These models were manually constructed and probability tables were created by manually filling in each cell in the table with a probability determined by the author. As such, these are also classified as human judgment models. The construction times for all measured models used for testing the second hypothesis are presented in figure 7-3.

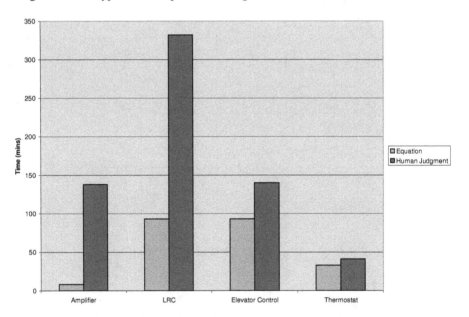

Figure 7-3

Human Judgment Model Construction Time Comparison

Hypothesis 2A is tested with the data of figure 7-3 resulting in a rejection of the null hypothesis at the 95% level of confidence. In the Amplifier, LRC and Wing models of figure 7-2, the use of nine volunteers to predict the outputs added significant time to the construction process. This can be seen in tables A-1, B-1 and C-1 of appendices A-C. The average time for each of the nine humans to make their predictions was 8.2 minutes, resulting in over an hour in each case to collect the human judgments. In the Elevator Control and Thermostat models where probabilities were generated from a single person, the time of model construction is very similar to the time required to build the equation-based model.

The results obtained in the test models and cases do not support a conclusion that there is an advantage to using human judgment Bayesian networks over equation-based models. Human judgment models generally demonstrate less accuracy than their equation-based equivalents. Construction of human judgment models requires about the same amount of time if the probabilities are entered manually into the node probability tables by a single person. Considerably more time is required if collecting predictions from multiple people. The one possible exception may be for very complex modeling problems in that the test data shows a lower time of construction as compared to the estimate while accuracy is better in one of the two cases. Investigating very complex models is beyond the scope, capability and budget of this investigation but should be considered for future research.

7.2 Comparison of Equation-Based Models with Formula Bayesian Networks

The comparison of accuracy uses the error measurements in the models of appendices A, B, E, F and K. The five models are used to generate nine cases for testing the first hypothesis. A summary of model errors is presented in figure 7-4. Hypothesis 1B is tested with the data of figure 7-4 resulting in a failure to reject the null hypothesis at the 95% level of confidence. The error associated with the formula Bayesian network models is higher in five of the nine cases and less in the other four cases as compared to the equation-based models. The averages for all nine cases are nearly the same for each modeling method. This is a logical result since both sets of models were built using the same set of equations.

90

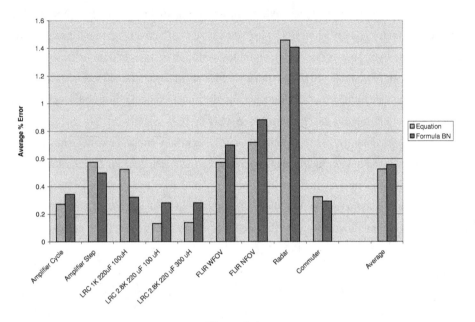

Figure 7-4

Formula Model Error Comparison

The comparison of time of construction uses the models of appendices A, B, E, F and K are presented in figure 7-5. Hypothesis 2B is tested with the data of figure 7-5 resulting in a rejection of the null hypothesis at the 95% level of confidence. As can be seen from figure 7-5, times of construction are very similar for both simple and complex models. The average over all five models is nearly the same number. This indicates that constructing models using equations takes roughly the same amount of time whether using The Extend modeling and simulation package or the Netica Bayesian network package.

The only interesting point of this comparison is the LRC Bayesian network formula model. The first attempt to build this model used intermediate variables of the equations found in appendix B the same way the equation-based model of figure B-2 of appendix B is constructed. This model failed to work because the equation is a second order equation which may have complex roots. When using intermediate variables in a Bayesian network, some of the

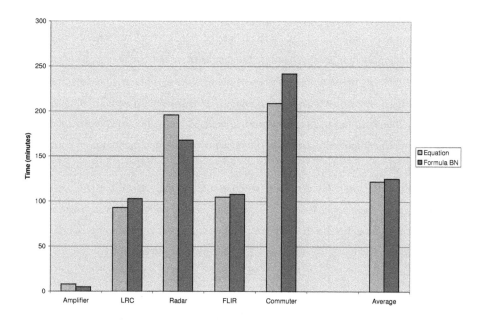

Figure 7-5

Formula BN Model Construction Time Comparison

variables were converted to complex numbers when the equation to table option was used due to the calculation of all possible combinations of the bins of each continuous variable. This resulted in some nodes having a mixture of real and imaginary numbers in the probability distribution. The formula network was reconfigured as shown in figure B-8 so that the entire equation for calculation of current was entered as one very large formula into node "I". Although this corrected the problem with imaginary numbers, it still resulted in a poor prediction of current and voltage out for the lowest time bin of 0.0 to 0.1 seconds in two of the three predictions. This anomaly can be seen in figures B-13 and B-14 of appendix B.

The results of this comparison demonstrate that equation-based modeling and Bayesian networks constructed using formulae with the chosen software packages produce equivalent results. Construction and comparison of more models is highly unlikely to change the results of the hypothesis testing. The advantage of using a Bayesian network is that its output is a probability distribution. An equation-based model would have to be run multiple times to get the same distribution. A disadvantage of using a Bayesian network constructed with the Netica software is that it may require the formula to be entered as one very long equation to prevent the generation of imaginary numbers. Although this results in a smaller network, it is more difficult to program and far more difficult to troubleshoot than breaking the equations down into a series of simpler equations using intermediate variables. This limitation is related to the particular software selected and not to Bayesian networks in general. Evolution of Bayesian network software to hierarchical modeling capability may improve the implementation of equations within the models.

7.3 Comparison of Equation-Based Models with Computer-Generated Bayesian Networks

Model errors from appendices A, B, and C are used to compare the accuracy of computer-generated models with the baseline. The three models are used to generate seven cases for testing the first hypothesis. A summary of model errors is presented in figure 7-6.

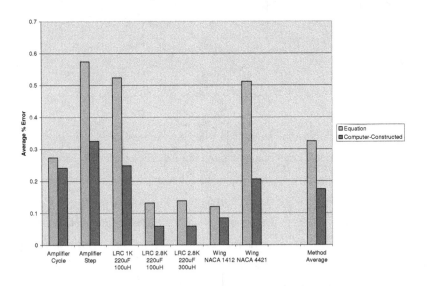

Figure 7-6

Computer-Generated BN Model Error Comparison

93

Hypothesis 1C is tested with the data of figure 7-6 resulting in a rejection of the null hypothesis at the 95% level of confidence. The error associated with computer constructed Bayesian network models is lower in all seven cases as compared with the equation-based models. The magnitude of the difference in error is great enough to establish a statistical difference between these two methods at 95% confidence.

Three cases are of particular interest in figure 7-6: Amplifier Step, LRC 1KΩ 220μF 100μH and NACA 4421 wing. In all three cases the computer-generated Bayesian network had substantially less error than the equation-based models. This can be traced to non-linearity in the test data of these three cases which are not captured by the equations. In the amplifier model where the step function is used, there is noise at the input and output of the amplifier at the lower end of the step when zero voltage is the input as can be seen in figures A-6, A-7 and A-19 of appendix A. In the LRC 1KΩ 220μF 100μH model the voltage does not decay to a steady state value near zero as predicted by the equation which can be seen in figures B-4 and B-12 of appendix B. This was traced to charge leakage from the capacitor. The voltage decays instead to a value of approximately 0.2 volts where the capacitor charge leakage and rate of charge reach equilibrium. The NACA 4421 has flow separation which occurs at higher angles of attacks resulting in a loss of lift. This occurs in thicker wings as can be seen in figures C-5 and C-12 of appendix C. The computer-generated Bayesian networks are able to learn the existence of these phenomena and accurately predict their presence in other similar configurations.

Improved accuracy was not expected prior to conducting the research. Early tests did show that Bayesian networks could learn relationships between variables including any non-linearities that are not captured by equations, and that this could result in improved accuracy. However, early tests also showed a loss of accuracy when discretizing continuous data sets into bins. It was theorized that improvement in one area would cancel out the additional error added in the other area during construction of a Bayesian network by a computer. Development of the derivative method of discretization, incorporation of the neural network to provide the ability to predict cases not represented in the data set, and replacement of Dirichlet probabilities with normal distributions of the data all improved the accuracy of the discretized models. These developments also allowed the data to be discretized into a larger number of bins increasing resolution of the final model. The total combination of factors resulted in very small errors introduced during the final discretization process. These small errors were vastly overwhelmed by the gains in accuracy achieved through learning of the variable relationships directly from the data. The accuracy improvement by itself presents an exceptionally strong case for considering the use of computer-constructed Bayesian network models.

The comparison of time of construction uses the models of appendices A, B, D, E, F and K. The construction times of these models are presented in figure 7-7.

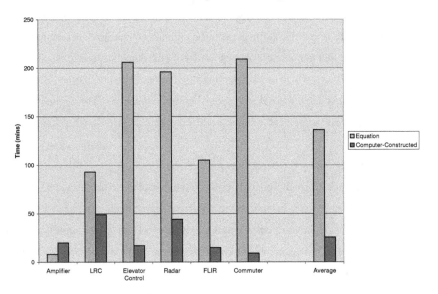

Figure 7-7

Computer-Generated Model Construction Time Comparison

Hypothesis 2C is tested with the data of figure 7-7 resulting in a failure to reject the null hypothesis at the 95% level of confidence. The computer-generated models take less time to construct as compared to equation-based models in five out of six cases. The difference is so great in these five cases that it supports a conclusion that construction time for computer-generated Bayesian networks is less than equation-based models at 95% confidence. This result was expected in that computers can perform certain tasks such as complex mathematical calculations and filling in large probability tables much more quickly than humans. This result further strengthens the case for considering the use of computer-generated Bayesian network models as the results support the research goal of reducing M&S costs.

95

All of the cases provide noteworthy observations. The amplifier model is the only comparison where the equation-based model took less time to construct as compared to the Bayesian network. This is caused by unique factors associated with this particular model. The equation model was extremely easy to construct as the modeling software package came with a pre-built amplifier element. The more complex circuit of figure A-1 is contained in the single block labeled "Amp" of figure A-2 in appendix A. The Bayesian network took longer to construct as the input and output data consists of two very different waveforms. This required the construction of two separate Bayesian network models and two neural networks; one to represent each waveform. Two models are required because of the limitations of the derivative method described in chapter 4.

The LRC and Radar models have the highest construction times for the Bayesian network models. As presented in table B-1 of appendix B, 41 of the 49 minutes needed to construct the LRC model were consumed by computer run time of the BN builder program. Although this network contains only five nodes, the input data set contains 808 rows while the program generated 303 additional rows of data to predict the three circuit configurations tested. By contrast, the Commuter model contains 30 nodes (29 continuous, 1 discrete). The input data set contains 25 rows of data and no additional data is generated. Computer run time is 3 minutes to construct this model with the BN Builder software. This demonstrates that model construction time is influenced much more by the number of rows of data including any neural network predictions than by the number of columns in the data set. This also demonstrates the efficiency of the derivative method where 29 continuous nodes are simultaneously discretized within a total network build time of about three minutes.

The Elevator Control model has the greatest difference between the two construction methods. This case is unique in that both models do exactly the same control function with the equation-based model using a rule technique and the Bayesian network using probabilities. The elevator of appendix D has six call buttons on the four floors, each with two possible conditions. The elevator can be moving up or down, has four destination selection buttons each with two conditions and can be at one of four floors when a decision must be made. A rule based approach must cover ($2^6 * 2 * 2^4 * 4^1$) or 8192 possible combinations. All combinations can be reduced to 512 rules to cover elevator movement. This is possible because multiple conditions can be covered by a single rule. For example, if the elevator is at a floor below two and the direction is up, one rule covers both the second floor destination select and the second floor up call button. Programming the rules in an equation-based model requires 111 function blocks as shown in figures D-1 through D-3 of appendix D. By contrast, the computer-generated model requires only 13 nodes as shown in figure D-5 of appendix D. Although the tables of the position and destination node are large, the time required to learn the probability tables is small for computer learning. The conclusion is that there is a great advantage to using Bayesian networks or influence diagrams for decision or control models as compared to a rule-based approach.

The exact type of network to be used is determined by the specific application. If the problem is rule-based where the desired result is for the network to provide a specific output for a given set of inputs, all that is required is a probabilistic inference. A Bayesian network will suffice in this application. An example would be a target decision support network where the requirement is to determine whether a contact is friendly or hostile as demonstrated in appendix L. Based on sensor inputs, this only requires probabilistic inference. If the problem is more complex and the application requires optimization based on conflicting goals, an influence diagram is used. Influence diagrams contain utility nodes allowing the assignment of utility values to outcomes of the decision. If the target network must now make a recommendation to either shoot down the target or to hold fire, that decision must be weighed with the utilities of the four possible outcomes: target friendly and held fire, target friendly and shot down, target hostile and held fire, target hostile and shot down. The recommended decision will now be the maximum value of the sum of the utilities assigned the outcomes of each decision times the probability the target is hostile or friendly as demonstrated in appendix L.

The LRC, FLIR and Commuter models are of interest in that the structural relationship found using the BN PowerConstructor program did not match the relationship of the equations derived for the same models. In the case of the LRC model, BN PowerConstructor found no relation between the value of the inductor (L) and any of the other nodes. This is because for the values of R, C, and L drawn from available parts on hand, the roots of the equation were real and the delay in voltage rise caused by the inductor was too short in duration to be measured with the test equipment. This demonstrates an important advantage to the use of Bayesian networks. If the learning data set bounds the upper and lower limits of the input variables, a Bayesian network may find a simpler solution if some variables do not influence the outcome in this particular region.

In the case of the FLIR and Commuter models, the BN PowerConstructor program found variables relationships that were not contained in the equations. For the particular FLIR system tested, a relationship was found between the display polarity and the maximum detection range. This occurred despite the fact there is no theoretical basis in infrared theory for such an effect to occur. Because of this the resulting computer model is far more accurate than either the equation-based model or formula Bayesian network as can be seen in figure F-15 of appendix F. Similarly, the Commuter model found relationships between both day of the week and time of departure and the total commute time as can be seen in figure K-14 of appendix K. This demonstrates another important advantage to using computer-generated Bayesian networks in that structural learning from data sets may find relationships not contained in the equations that can significantly influence the outcome of the model or simulation.

7.4 Discussion and Conclusions

The comparison of modeling methods provides insights into current methods of modeling and simulation. Most models and simulations constructed today for engineering trade studies use an equation-based approach. Therefore, this type of model was used as a baseline for this research. Comparison of this method to manually constructed human judgment problems found no advantages in that these Bayesian networks on average took longer to construct and were less accurate as compared to the baseline. Comparison of formula Bayesian networks to the baseline found no compelling reason to change modeling methods. These two methods were essentially equivalent in accuracy and time of construction while the formula Bayesian networks experienced some problems with imaginary numbers because of the software package used for the comparison. These first two methods of Bayesian network construction are currently what are available in commercial software packages. These test results may explain why there has been no move towards wider use of Bayesian networks in engineering applications.

The results of this research demonstrate that there is a two compelling reasons to consider the use of computer constructed Bayesian networks in engineering model construction. Research results demonstrate that this method is statistically both faster in terms of model construction time and is more accurate than using present equation-based methods. Figure 7-8 presents a comparison of construction time and model complexity.

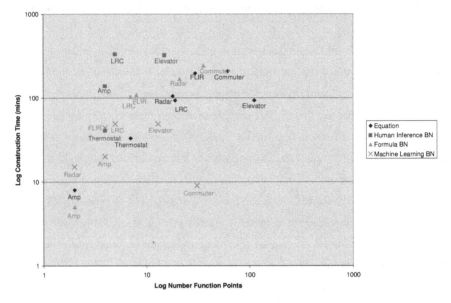

Figure 7-8

Model Time-Complexity Comparison

Figure 7-8 shows a log-log plot of the number of function points versus the time of construction. A function point is defined as one functional element in the equation-based Extend modeling package or one node in the Netica Bayesian network software package. One observation is that for equivalent models, the Bayesian network models require fewer function points than the equation models. This is possible because equation-based models often require multiple constants and calculation of intermediate variables to reach the final answer. Computer constructed Bayesian networks do not need constants and can find relationships directly between input and output variables.

A second observation is the relationship between time and complexity. Excluding the elevator control which was a rule-based controller as opposed to an engineering model, a linear relation can be seen between complexity and time of construction for both the equation models and the formula Bayesian network in figure 7-8. This indicates that when using equations in either modeling package, that as the complexity of the modeling problem increases the time of construction increases exponentially. By comparison, there is no observable relationship

99

between complexity and time of construction for the human judgment or computer constructed Bayesian networks. As previously described, the time of construction of computer-generated models is most heavily influenced by the size of the database and the amount of data generated by the neural network. As models become more complex, more data is required to learn the relationships requiring more time to construct the model. However, the rate of increase of time with complexity for computer-generated models is less than that for equation-based models. This leads to a conclusion that the more complex the modeling problem, the greater the construction time savings the computer-generated Bayesian network will have with respect to an equation model.

The cost of modeling and simulation is driven mostly by the human labor involved in the process. Although computer equipment and software require upfront investments, the cost of computer run time once purchased is negligible. Additionally, all demonstrations in this research have used commercial off-the-shelf equipment, software products and software development packages. The average times to perform specific tasks while constructing the models of appendices A, B and F are presented in figures 7-9 and 7-10.

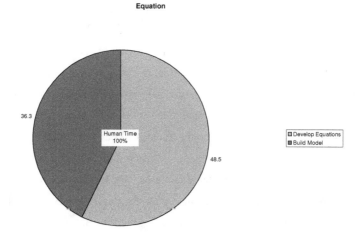

Figure 7-9

Equation Model Construction Time Task Percentages

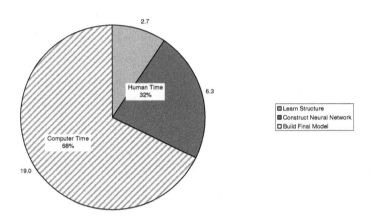

Figure 7-10

Computer BN Model Construction Time Task Percentages

As can be seen from the figure, not only has the average time of construction been reduced from 69 to 28 minutes, but the task loading requiring human work has been reduced from 100% in the equation-based model to 32% for the computer-generated Bayesian network. Reviewing the breakdown of model task times contained in the appendices, as model complexity increases, total construction time increases. However, the human tasks associated with model construction remain nearly constant. Learning the network structure and constructing the neural network both require human input, but are computer-aided tasks. The increase in construction time is almost completely attributable to increased computer run time of the BN Builder program. This leads to a conclusion that models created using computer generated Bayesian networks would be much less expensive to build than equation-based models. Not only is time of construction significantly less, but the human labor involved is also reduced. Because there is no longer a strong relationship between complexity and human labor required, costs to construct computer-generated Bayesian networks are not sensitive to problem complexity.

Tests results show that a computer-generated Bayesian network is not superior to an equation-based model in every case. Each modeling method has certain advantages depending on specific circumstances of what is being modeled. Based on test results, the following circumstances favor the use of an equation-based approach:

- Validated equation-based models already exist
- Modeling function blocks already exist
- There is a scarcity of available data on what is being modeled
- The element being modeled does not require many function points
- Inputs and/or outputs are very different functions such as linear and non-linear data

The following circumstances favor a computer-generated Bayesian network approach

- Database of observed or test data already exists
- Problem is not well understood and/or equations do not exist
- Problem is complex
- Input and output functions are similar
- There may be unknown non-linearities
- Hidden variable relationships may exist
- Problem is a control application or decision problem

The conditions most favorable to computer-generated models are those with the greatest potential to reduce the time of construction and expense of modeling and simulation. Building an equation-based model is straight forward if equations exist and are readily available. It is the complex problems where equations do not exist or non-linear elements are present that make it difficult and time consuming to define the problem using an equation-based approach. The results of this research indicate that it is now possible to rapidly construct models of very complex systems. Because cost is primarily driven by human labor required to construct models, this reduction in time and transfer of labor from human to computer should translate into greatly reduced cost. The goals of the original research have been exceeded by not only demonstrating that Bayesian network engineering models can be created in less time at reduced cost, but that these models are more accurate than their equation-based equivalents.

8. Integrating Equation-Based Models and Bayesian Networks

Research results comparing equation-based models with Bayesian networks conclude that there is no single method that works best in all models or simulations. Each method has advantages and disadvantages depending on the specific circumstances of the problem. Equation-based models and Bayesian networks are not mutually exclusive methods of modeling and simulation. Based on the software development described in chapter 6.2, the two methods can now be used simultaneously within the same environment. When modeling complex systems, the problem is usually broken down into smaller, simpler subsystems that are constructed, tested and then integrated into the final complex model or simulation. This approach lends itself to creation of integrated models where each component to be modeled is individually evaluated to determine which modeling method would be best under the particular circumstances. Appendices D, H, I, J, L and M demonstrate the utility of this approach with integrated models that contain both equation-based elements and Bayesian networks working together to create a single simulation.

Another problem in modeling and simulation has been how to mix models that have different levels of fidelity and resolution. Chapter 3 presents an approach to multi-resolution in which inputs to the model are represented as random variables, probability distributions are defined over their range of potential values, and Monte Carlo simulation is used to create multiple samples and calculate a final output distribution. Because Bayesian networks are distributions, this method works well in an integrated modeling environment. Appendices L and M demonstrate this approach by integrating Bayesian network subsystems that are of different fidelity and resolution from other equation-based sub-models.

8.1 Bayesian Networks as Subsystems of Equation-Based Models

Appendix L demonstrates the integration of the equation-based F-16 radar model of appendix E with the Bayesian network B-26 radar cross section model of appendix G. A second radar cross section of a 1/15 scale model of a Boeing 737 commercial aircraft is also constructed from unclassified test data. These models were constructed and validated from two entirely different sets of measured data using two different methods. As such, these models have completely different fidelities and resolutions.

To simulate a radar tracking engagement, an equation-based motion model is added to the other models. The motion model allows the aircraft to fly across the viewing angle of the radar. As the aircraft moves, both the range and aspect angle of the target to the radar changes between each radar sweep. Example simulation tracks of each aircraft are shown in figure 8-1.

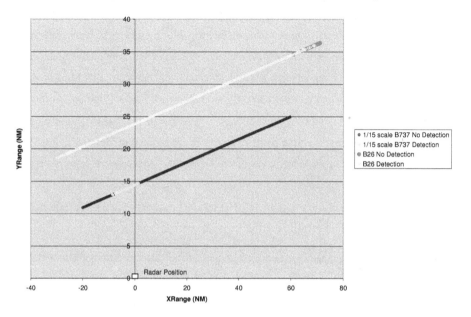

Figure 8-1

Radar Tracking Simulations

The radar begins to detect the B-26 just inside the maximum radar display range of 80 NM. The real world phenomenon of target scintillation, where the target returns fade in and out near the maximum detection range, is observed on the B-26 track. Once the B-26 moves closer to the radar, the return is so strong that a continuous track is maintained even as the aspect changes. The much smaller model of the Boeing 737 aircraft is harder to detect. The target is visible on the display for only a short time. In this case, the target is detected not as a function of range, but of changing aspect angle. The radar detects the aircraft as the target aspect shifts to the left side of the aircraft where a higher radar cross section exists as shown in figures L-5 and L-7 of appendix L. The target detection is lost as the aspect angle moves to the rear quarter which has a much lower cross section even though the target continues to get closer to the radar. Although it is impossible to obtain test data to validate this integrated simulation, the results match the real world experience of the author in operating air-to-air pulse radars.

Appendix M provides an example of an engineering trade application where a system with a validated equation-based model exists. This example uses a small robotic vehicle that can be easily modified so that model predictions can be compared to test data. The problem looks at replacement of the two electric motors with a different pair of motors. The problem is to predict the speed of the vehicle system with the new motors. Instead of building a new equation-based model of the motor, a Bayesian network model is created from test data of RPM of the motor at various torque loads and power settings. The Bayesian network model is then substituted for the existing motor model in the robotic vehicle model. The test vehicle was also modified with the new motors to compare actual vehicle performance to the integrated model prediction. Due to a difference in physical size, it was also necessary to change the drive system from gears to a belt and pulley system. These parameters were entered into the model so that the model parameters matched the configuration of the test vehicle. The test technique for testing the motors and the vehicle system were different resulting in the Bayesian network motor model having a different fidelity and resolution than the equation-based system model.

The equation-based model and the Bayesian network model were both used to predict the performance of the car under four different sets of conditions. The mean of the prediction distribution is compared to the mean of the test data for both the baseline equation-based model and the modified integrated model with the error shown in figure 8-2. As can be seen in figure 8-2, the average error over the four conditions tested is approximately the same. As shown in appendix M, the distributions of the test samples all fell within the limits of the predictive distributions created by the models. The average error of approximately 5% is within the accuracy of the test methods used to collect the performance data.

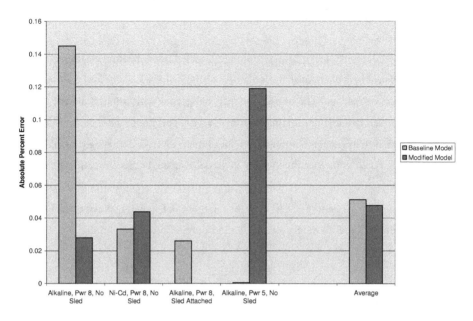

Figure 8-2

Robotic Vehicle Model Prediction Error

8.2 Bayesian Networks/Influence Diagrams Controlling Equation-Based Models

Appendix D provides an example where a Bayesian network is used to control an elevator. Although the research of chapter 6 showed that there was a strong advantage to constructing control elements using Bayesian networks, it would be very difficult to accurately model the physical components of the elevator with a Bayesian network. The elevator simulation includes a random people generator that assigns passengers to different floors and simulated desires to move to a different floor. The elevator itself is constructed with measured delay times for movement, door open and close and maximum capacity based on measurements from the elevator in the Science and Technology Building II at George Mason University. Additionally, the Extend M&S package has an animation capability allowing visualization of elevator operation. All these factors support use of an equation-based approach to construct a model of the elevator. Research showed that using a rule-based approach with an equation model for the control logic is difficult and time consuming to construct. An integrated simulation in which control is provided by the Bayesian network and the other components of the elevator use an equation-based approach provided a superior solution with respect to speed of construction and simplicity.

Appendix J demonstrates a similar simulation with a home heating system. The baseline house with all its heat loss mechanisms such as the walls, windows, doors and roof are modeled with equation-based elements. The heating system is also an equation-based model. The model is modified by replacing the single setting thermostat with a programmable thermostat that allows the temperature to change automatically four times per day. The programmable thermostat, which controls the heating system, is modeled using both an equation-based model and a Bayesian network. The outside temperature is varied over a 24 hour period using temperature data for the Washington DC area for the month of January. Each model is run using a Monte Carlo simulation with the results shown in figure 8-3.

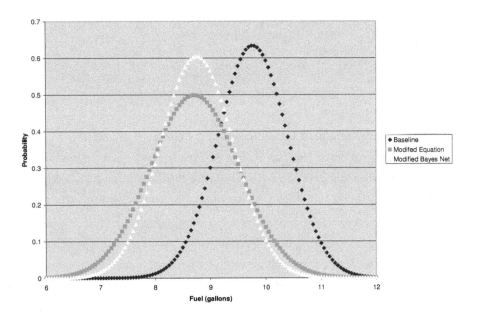

Figure 8-3

Home Heating System Simulation

The programmable thermostats result in an average decrease in fuel consumption of approximately one gallon per day as compared to the baseline fixed temperature thermostat. Because the thermostat was a simple model requiring very few elements to construct, both models had similar outputs and took about the same amount of time to construct.

Appendix H demonstrates the use of modeling and simulation in business process reengineering. This example looks at screening of loan applications for approval or disapproval. The baseline process uses two humans to screen loan applications to determine whether to approve them or route them to a second, more thorough review for final determination. The baseline model is modified so that this initial screening process is replaced by a computer screening that conducts the initial review. The modified model uses a rule-based approach to route the loan applications based on a simple set of screening rules. The simplified rules identify any application that fails to meet a specified threshold value in several areas and routes any flagged applications for the second review by a human. A Bayesian network model is also created that uses a file of previous loan cases to calculate the node probabilities. The Bayesian network is implemented so that it routes applications whose attributes indicated less than a 75% probability of repayment for the second, human review. The results of running the baseline model as compared to the two modified versions of the model over a 40 hour work period are presented in the simulations of figure 8-4. The rule-based screening process resulted in a process that was 18% less productive than the baseline human screening process. By contrast, the Bayesian network model resulted in a 103% improvement in productivity. This simulation demonstrates the importance of the use of modeling and simulation when considering process changes. Although one would expect the introduction of automation to improve the productivity of a process, this example shows that this is not always the case. In this case, complex interactions caused unintended consequences as described in appendix H.

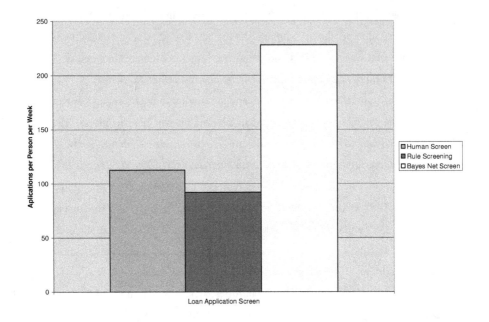

Figure 8-4

Loan Application Screening Simulation

Appendix I is also a business process simulation. The baseline simulation is a virtual representation of the electrical repair shop at an automotive center. The baseline simulation routes cars to one of three mechanics using a first in, first out routing process. The simulation is modified by recording reported symptoms that owners report when dropping off their cars for repair. Not all reports are accurate. The modified model uses a rule-based approach to route the cars based on a simple set of diagnostic rules that evaluate the reported symptoms. Cars are then routed to the mechanic best suited to make the repair of the fault based on the diagnosis. The model was modified again by replacing the rule-based procedures with a Bayesian network to diagnose the most likely problem. Probability distributions for the network were learned from previous car repair cases. Again, the cars are routed to the mechanics best suited to make the repair. The average weekly gross income for all three simulations is presented in figure 8-5.

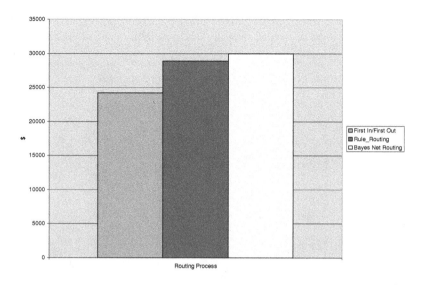

Figure 8-5

Car Electrical Repair Simulation Gross Weekly Revenue

Both of the diagnostic routing procedures demonstrated improved performance over the baseline simulation. The Bayesian network routing demonstrated a higher increase in weekly revenue with a 24% improvement as compared to the rule-based routing at 19%.

Appendix L uses an influence diagram as part of an air defense network to decide whether to fire a weapon in response to a radar contact. A Bayesian network created using human judgment for node probabilities was modified into an influence diagram by the addition of a utility node and a decision node. The utilities were set so that optimal decisions were made by firing at targets determined to be hostile while not firing at neutral or friendly targets. The network was tested by integrating it with the radar/radar cross section model described in section 8.2 and as shown in figure L-11 of appendix L. The equation-based model determined some of the node states based on the motion and target aspect of aircraft. Other states were generated by random number generators set to likely values of the type of aircraft. Random errors and missing values were also added to the inputs. A second version of the baseline influence diagram was created by adding arcs from the node "Identity" to nodes "EW" and "Kinematics".

This was done to allow the network to include the airspeed of the target and whether the radar was on in determining if the contact was hostile. The probabilities and utilities were then learned from the simulation.

A comparison of the two simulations is presented in figure 8-6.

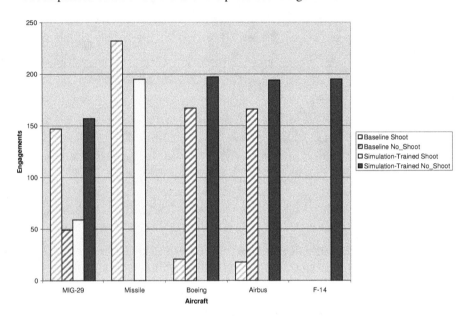

Figure 8-6

Air Defense Test Results

As can be seen from the test results of figure 8-6, the simulation-trained network provided much better decisions and target identifications than the human-judgment baseline. The baseline network correctly identified and recommended firing at all incoming missiles, but also recommended firing at 34% of friendly and 10% of neutral aircraft. It also recommended shooting at 75% of MIG-29 aircraft based solely on aircraft type. The test results demonstrated that the network had difficulty in determining the type of aircraft and a flaw in logic by using aircraft type as the primary determinate of whether it was hostile. Additional arcs were added to change the logic to observe aircraft actions in determining hostile intent. A decision policy was implemented that non-friendly aircraft flying towards the radar position at speeds greater than

500 knots with their targeting radar on were considered hostile. By learning the probabilities and utilities from the simulation, the simulation-trained network provided much better decisions during tests recommending firing only at missiles and MIG-29s demonstrating hostile intent. Provided that the simulation is an accurate representation of the real world, this example demonstrates a method for accurately training and testing of very complex influence diagrams.

8.3 Discussion

The six examples presented above demonstrate that Bayesian networks and equation-based models can be used together to form an integrated simulation methodology. Because there is no single method that is optimal in all circumstances, the integration of both types of models allows the model builder to choose different approaches for different subsystems, selecting the best approach for each subsystem. The previously demonstrated capability to rapidly and accurately build Bayesian network models from data should allow the construction of new, complex models and simulations that are not feasible using a single method. The ability to build rapid and inexpensive Bayesian network models also improves the capabilities of the modeling and simulation community to quickly conduct trade studies to determine if a proposed change to an existing system results in the desired outcome.

The radar tracking and robotic vehicle examples demonstrate that this technique can accommodate the requirement for constructing mixed-fidelity, mixed resolution models. Measurement error in the data used to construct the Bayesian network will be represented by a higher spread of values in a constructed Bayesian network. The resolution of a Bayesian network is primarily a function of the number of bins used for continuous variables. A Bayesian network constructed from a data set contains both a central tendency and spread that incorporates uncertainties from both the fidelity of the data and the resolution of the network. Using this data with a Monte Carlo method in the equation models allows the calculation of probability distribution as the output. This output provides not only a most likely answer, but the range of possible values that could possibly occur along with a probability for each possible range of answers. This approach provides a more complete solution to an engineering trade study than conventional sensitivity analysis.

The research also demonstrated the use of complex simulations to train influence diagrams. This approach allowed the network to test the outcomes of the decision options for different combinations of inputs. The network learned the probabilities and utilities from the results of each simulation. After many simulations, the network learned which decisions resulted in the most favorable outcomes. Provided the simulation is an accurate representation of the real world, the trained and tested influence diagram can then be used as either a decision aid or autonomous decision system. This approach may be particularly useful in future unmanned vehicle control systems where a high degree of autonomy is required. The examples provided above demonstrate the feasibility and utility of integrating Bayesian networks and equation-based models. The exploitation of this new capability should lead to multiple follow-on research projects. This approach may prove fruitful in a number of scientific disciplines that make use of models and simulations.

9. Conclusions and Future Research

Background research found that modeling and simulation is an important tool in the field of Systems Engineering. This tool is one of the few that has demonstrated the simultaneous achievement of bringing a better product to market in less time at a lower cost. These benefits apply to both large and small development efforts. Unfortunately, the current high cost of modeling and simulation puts this tool out of reach of many small development efforts. This is the result of current practice which uses an equation-based approach to modeling and simulation. An equation-based model is only as good as the equations that make it up. It is extremely difficult in complex system modeling to capture all the possible variables that may influence a problem and to define the relations between the variables using equations. As model builders attempt to create high-fidelity models by searching for and adding additional variables and equations, the costs continue to spiral upward.

The approach taken in this research was to explore the use of Bayesian network models as alternatives to equation-based models. Bayesian networks have an advantage as they define relations between variables in terms of conditional probabilities. Where equations-based models must use Monte Carlo sampling to generate a probability distribution over multiple computer runs, the Bayesian network can calculate these distributions in a single cycle. The use of probability distributions not only provides more complete answer to a problem by specifying both a central tendency and range for the answer, it also allows for the integration of models of mixed fidelity and resolution. The major drawback of Bayesian networks is the lack of good discretization techniques for continuous variables. A second drawback is that if the probabilities were learned from an input data set, the completed network can only respond to discrete queries for data that were provided in the input learning set.

9.1 Improvements to Discretization

The first contribution of this research is to propose a new, multivariate approach for discretization of continuous variables. This method was tested against four iterative search and score methods. Results showed that the derivative method performed as well as the other four methods and was faster in all cases. Based on this limited sample of discretization methods and data sets, a conclusion can not be made that this method is superior to any other discretization method. Further research should be conducted to better quantify the results. However, the

114

method was much faster than any method evaluated and did not suffer from memory limitation problems experienced by the software implementing the iterative methods.

There are a number of other potential follow-on research projects that should also be considered. Metrics should be developed that allow the algorithm to determine an optimal number of cut points. Iterative methods should be applied to the cut points determined by the derivative method to see if a better discretization can be produced in less time than starting with random cut points. The derivative method should also be applied to iterative discretization and structural learning to see if better structures can be obtained by alternately updating the discretization and structure between iterations until the two converge.

The research also addressed limitations in current Bayesian network software packages where Dirichlet learning is used to learn the probability tables from data sets. This limitation can cause problems in continuous nodes with a large number of bins in that there may be residual probabilities in the cells which had no data values. These probabilities, especially when the true answer lies near either end of the range of bins, can skew the central tendency and spread of the answer. The research proposed to address this issue by substituting normal probabilities of the data in place of the Dirichlet probabilities in the continuous leaf nodes. This approach provides an effective method of building high resolution Bayesian network models without loss of accuracy. It also provides a method to incorporate additional uncertainty when predictions were added to the input data set.

The implementation for this research was limited to continuous leaf nodes and normal probabilities. This was all that was required as the examples tested were engineering problems where the outputs represented in the leaf nodes were the variables of interest. The data sets were all expected to be normal distributions. This is obviously not the case in all problems. Future research should look at potential improvements to network performance if all continuous nodes are converted to probability distributions. It should also look at determining what type of distribution best fits the data and use a distribution of that type in the continuous node probability tables.

9.2 Predicting Values for Incomplete Data Sets

The second contribution of the research is the integration of neural networks into a Bayesian network construction algorithm. This integration provides a number of advantages for Bayesian network construction. The neural network can produce additional data for the input training set by predicting missing values or providing more data in areas that contain sparse observations. The implementation allows the calculation and inclusion of uncertainties associated with the predictions to be included in the final Bayesian network probability tables. The result is an output Bayesian network that can reason with inputs not provided in the input data set as long as the inputs are bounded by the input data. The use of a neural network also acts to smooth scattered data sets insuring good performance of the derivative method discretization of continuous variables.

The neural networks are currently constructed and trained as a separate process with the trained network used as an input to the Bayesian network build process. Further research should be conducted into algorithms for construction of neural networks from data sets. As software incorporates this capability, future research should look at computer constructed and trained neural network models. Automation of the neural network construction and training process would further reduce the amount of human task time required to construct computer-generated Bayesian networks. Additionally, the current software allows only a single neural network as input. If construction of neural networks can be accomplished by computer, the software should allow the generation and use of multiple neural networks to predict missing values in any number of columns.

9.3 Alternatives to Equation-Based Models

The primary focus of the research was to compare three methods of Bayesian network construction against equation-based models used as the baseline. The first comparison looked at manually constructed Bayesian network models using human judgment for the probabilities. The results of this comparison showed that Bayesian network models created using this method on average demonstrated lower accuracy and higher time of construction than the baseline models. The research used 9 participants which added considerable time to the process to collect and analyze the predictions. If using a single expert, the time to construct this type of model was about the same as the baseline. The results of the individual predictions did not correlate well with the confidence factor assigned by the participants. Even though some of the participants' predictions were comparable in accuracy to the baseline, there was no method to determine which one was the best prediction. Additionally, the participants with the best prediction for one problem did not have the best among all problems or even within different configurations of the same problem.

The second comparison looked at Bayesian network models manually created using formulae with the baseline. The results for both time and accuracy were very similar. Because the same formulae were used, this comparison showed that there is no advantage to using a Bayesian network software package to implement an equation as opposed to an object modeling and simulation software package based on the two software packages chosen. The Bayesian network model showed an advantage in calculating a more complete probability distribution as compared to Monte Carlo sampling with an equation-based model when there were a large number of variables in the model. The Netica Bayesian network software package did show some limitations with probability distributions of complex numbers.

The final comparison looked at computer-generated Bayesian network models constructed with software and algorithms specifically developed for this research. Time of model construction was compared resulting in a conclusion that computer-generated models can be constructed more quickly than the baseline with 95% confidence. This comparison assumes that the input data set is available for the learning of the structure and probabilities of the model. A comparison of the two methods was also conducted to compare model accuracy. The computer-generated Bayesian networks were constructed from an input data set that contained no observations for the combination of input parameters used to evaluate the accuracy. The comparison resulted in a conclusion that the computer-generated Bayesian network models were more accurate than the baseline with 95% accuracy.

This result was unexpected, as it was anticipated that the Bayesian network models would loose some accuracy during discretization of continuous data. Studying the models, it appears that the combination of research work associated with derivative method discretization, neural network integration and substitution of normal probabilities into the node probability tables combined to allow creation of high resolution Bayesian networks. These networks had very little loss in accuracy associated with disretization of continuous data. The Bayesian networks were able to learn the presence of non-linearities and unmodeled variables and/or relations from the data set. The networks were then able to apply these factors to different combinations of input values resulting in improved accuracy as compared to the baseline.

The comparison also looked at the relationship between time of construction and model complexity. The equation-based models showed an exponential increase in construction time with model complexity. The computer-generated models did not show a strong correlation of time with complexity. Computer-generated model construction times were most strongly correlated with the number of observations in the data set. Although more complex modeling problems will likely have larger data sets, this increase in construction time was almost solely attributable to longer computer run times. There was little increase in human tasks associated with larger data sets or more complex modeling problems. All problems evaluated were run on a commercial, off-the-shelf desktop computer, requiring no special equipment to construct the networks. It is therefore concluded that computer-constructed Bayesian network models will reduce the cost of constructing models and simulations by simultaneously decreasing the time of construction and the human labor required for construction. Further research is needed to quantify the savings. Demonstration of a faster, less labor intensive method of model construction that is relatively insensitive to complexity while simultaneously improving model accuracy is the third contribution of this research.

9.4 Challenges to the Conclusions

One may challenge the conclusions based on the assumptions applied to the research problem. The baseline equation models were constructed from formulae published in subject matter textbooks on the subject being modeled. The equation-based models were not able to take advantage of the test data that were used to create the computer-generated Bayesian networks under these conditions. There is no argument that if the data were used, further tests were conducted and additional variables and equations were implemented in the equation-based models, their accuracy would improve and in most cases could probably be equal to that of the computer generated Bayesian networks. Research indicates that this is current practice. This approach is trading additional time to improve accuracy. However, studies have also shown that this is precisely what drives up the cost of modeling and simulation development. Such an approach only strengthens the argument that computer-generated Bayesian networks should be considered as an alternative to equation-based models. The counter to this challenge is to ask why one would spend so much time and money achieving equivalent accuracy when computer-generated Bayesian networks allow the same accuracy to be achieved so much faster and presumably at greatly reduced cost.

A second challenge may be to the assumption that test data is available to create the computer-generated Bayesian network. Best practices for modeling and simulation, no matter what method is used for construction, requires that a set of test data be obtained and used to validate the model or simulation prior to use. This validation data must cover the limits within which the model is used for predictions. This is identical to the assumption used to construct the computer-generated Bayesian networks. In this research, the time of data collection was unavailable in most cases as the test data were obtained from outside sources. However, since it is required for any method used, it would be an equal addition to the times of any method. Although the percentages and confidence may change for how fast one method is as compared to another, the end result that computer generated models can be generated more quickly would not be different. The results of the model comparison also found that neither method was optimal in all situations. Lack of good data with which to construct a Bayesian network model is one consideration that may result in an equation-based model being the best choice for a particular application. A computer-generated Bayesian network model is not the best choice in every case.

9.5 Integrated Bayesian Network/Equation-Based Modeling

Because a Bayesian network model is not the best solution in every case, the research makes a fourth contribution by demonstrating that Bayesian networks can be integrated with equation-based models and simulations. This approach provides improved flexibility for model builders. When constructing models of complex systems, these systems are usually broken down into subsystems and components which are modeled and then integrated together. By demonstrating that Bayesian networks can be integrated with equation-based modeling and simulation, model builders can now pick and choose the best method to use for construction of lower level subsystems and elements. This will not only provide for more flexibility, but may allow for simulation of systems that can not be done using current rule-based approaches.

The research also demonstrated that equation-based simulations can be used to train influence diagrams. Through this process, the inputs and decisions are randomly generated and utilities based on the outputs of the simulation are calculated. The influence diagram node probabilities are updated and the utility values are set for the given decision at the end of each simulation. After multiple simulations, the influence diagram learns which decisions result in the best outcomes for a given set of input states. Provided the environment of the equation-based simulation is an accurate representation of the real world, this provides an improved method for the training of complex influence diagrams.

9.6 Summary

This research has developed the necessary methods to construct mixed discrete and continuous data Bayesian networks from data sets. Integration of neural networks allows prediction of additional data values expanding and smoothing the input data set. The result is the ability to construct a Bayesian network that can make probabilistic inferences to input conditions not in the input data set. In demonstrations on multiple modeling problems varying from simple elements to complex systems, these Bayesian network models outperformed their equation-based equivalents in both accuracy and time of construction. Because neither method was optimal in every case, the research demonstrated that Bayesian network models and influence diagrams could be integrated with equation-based models. The end result is an expansion of capabilities that should provide increased flexibility for the model builder and reduced cost to the model consumer. The results of this research are expected to result in the wider use of modeling and simulation in product development. It is also expected to benefit many areas of science and engineering where models and simulations are essential tools of the trade.

Appendix A - Amplifier Models

This set of models is a virtual representation of an electrical amplifier with gain set to a value of two. An electrical amplifier takes electrical signals as inputs, and acts as a multiplier where the output is the input signal times the gain of the amplifier. The electrical wiring schematic of an amplifier circuit is provided in figure A-1.

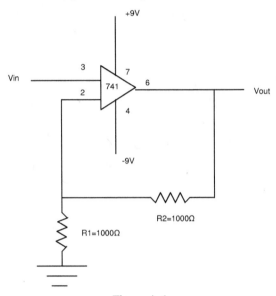

Figure A-1

Amplifier Electrical Circuit Schematic

The gain of the amplifier is determined by the equation:

$$Gain = (1+R2) / R1$$

The minimum and maximum output values of the amplifier are limited by the input voltages that power the amplifier; ±9V in the example of figure A-1.

Test data were obtained by constructing the circuit of figure A-1 and using two types of input waveforms: a step waveform and a cyclic waveform. The step waveform was created by switching a voltage source either on and off or back and forth between values at periodic time intervals. The cyclic waveform was created by switching a voltage source on and off or back and forth between values to charge the circuit of appendix B. The output of this circuit was then used as the input to the amplifier. Test data were sampled using a Lego® RCX computer coupled with a high accuracy voltage measuring interface from Lego Dacta®. Data were captured using LabVIEW® Software from National Instruments. The number of samples was determined by using the Monte Carlo sampling technique described in section 2.1 for a width of a 0.1 volts and 95% confidence.

A.1 Equation-Based Model

The Extend equation model of this circuit is provided in figure A-2.

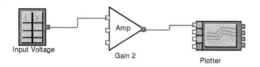

Figure A-2

Amplifier Equation Model

As can be seen in figure A-2, an amplifier component was available in the modeling package eliminating the need to construct the circuit of figure A-1. Equation-based modeling is considerably easier and faster if modeling elements already exist within libraries as compared to if they must be modeled from lower level elements.

A.2 Human Judgment Bayesian Network

The manually constructed Bayesian network model is provided in figure A-3.

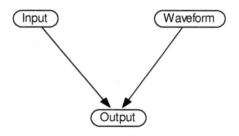

Figure A-3

Amplifier Human Judgment Bayesian Network Model

The probability tables were created through multiple predictions from nine volunteers who were provided with the test data shown in figures A-4 through A-8.

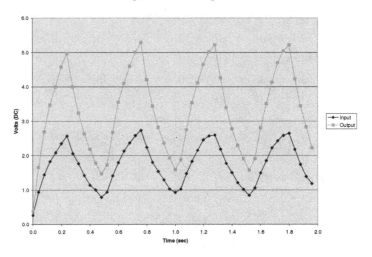

Figure A-4

Amplifier Cycle 1 Test Data

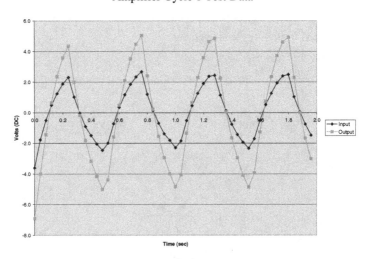

Figure A-5

Amplifier Cycle 2 Test Data

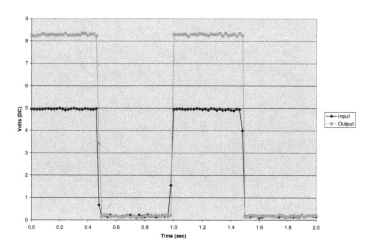

Figure A-6

Amplifier Step 1 Test Data

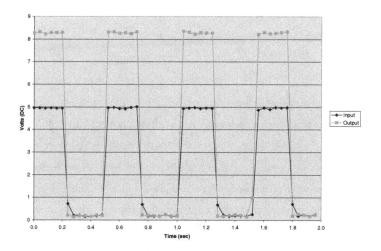

Figure A-7

Amplifier Step 2 Test Data

124

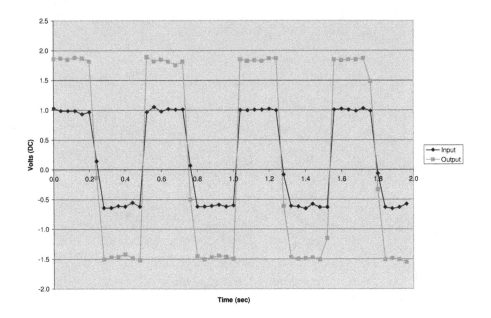

Figure A-8

Amplifier Step 3 Test Data

The volunteers were then provided inputs shown in figures A-9 and A-10. These two samples were chosen so as to sample one of each type of waveform shown to the volunteers. As previously described, the waveform generators were also circuits which were made from commercial electrical components available from retail electronic supply outlets. The inductors, capacitors and resistors were purchased in standard, mixed lots of values. Components for the input circuits were randomly selected from these standard values. All measurements were taken over a two minute period. Time constants for switching of voltages were obtained by dividing the time period by random selections of even integers so that complete cycles were measured.

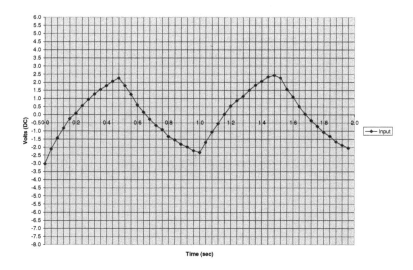

Figure A-9

Amplifier Cycle Input Voltage

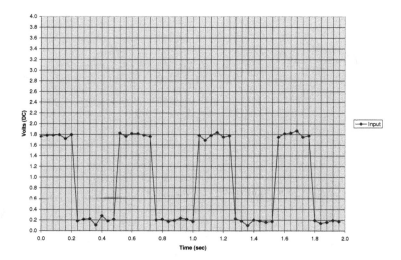

Figure A-10

Amplifier Step Input Voltage

126

The volunteers provided the predictions shown in figures A-11 and A-12. Test data, which the volunteers did not see, is provided for comparison.

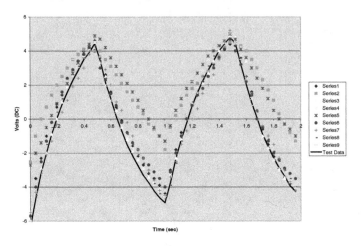

Figure A-11

Amplifier Cycle Output Predictions

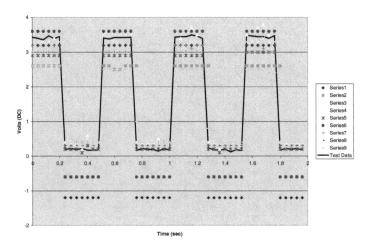

Figure A-12

Amplifier Step Output Predictions

Participants were also asked to provide a confidence assessment between one and five for their prediction with five assessing very high confidence and one assessing very little confidence in the prediction. These confidence assessments were used as a weighting factor during learning of probabilities. For each individual, the prediction was counted the confidence assessment times one hundred during learning. This results in a final probability table for the node "Output" where those who were more confident in their predictions had a greater impact on the final distribution than those who were not as confident. Confidence assigned was not an accurate predictor of accuracy. A comparison of the error with confidence score is presented in figure A-13.

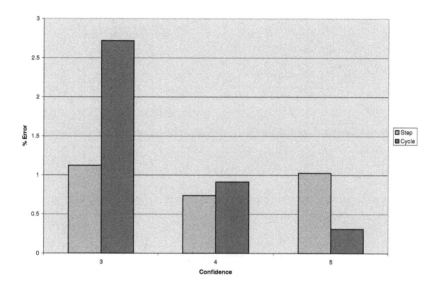

Figure A-13

Amplifier Error versus Confidence

As can be seen from figures A-11 and A-12, some of the predictions were very accurate. However, there was no way to predict which score was the most accurate.

A.3 Formula Bayesian Network

A second Bayesian network was created using the same formula as was used in the equation-based model. This model is shown in figure A-14.

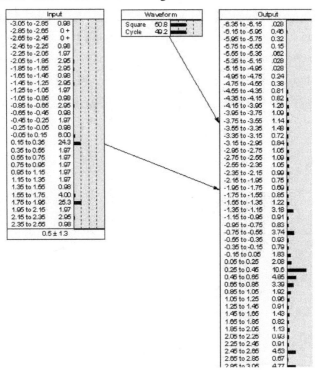

Figure A-14

Amplifier Formula Bayesian Network Model

This network uses a uniform probability for the variable "Input" and the equations to table option to calculate the "Output" probability distribution.

A.4 Computer-Constructed Bayesian Network

The final model created using the computer software described in section 6.1 is shown in figure A-15. The relation was manually constructed since there are only two nodes in the network.

Figure A-15

Amplifier Computer Network Structure

For the computer-generated networks, two separate models were created for the two different input functions. This was done because a node can have only one discretization and tests conducted during development of the derivative method showed poor results when radically different data sets were discretized for the same node. This condition is present in that the cycle input and output are linear functions while the step input and output are non-linear.

A neural net model was constructed for each waveform. The neural network is a fully-connected hybrid network containing two hidden layers with a sigmoid transfer function in the first and output layers and a Gaussian transfer function in the second layer. The network structure was the same for both waveforms and is presented in figure A-16. The cycle network was trained using the average data shown in figures A-4 and A-5. This consisted of a data set with 100 observations for the cyclic network and 157 observations for the step network.

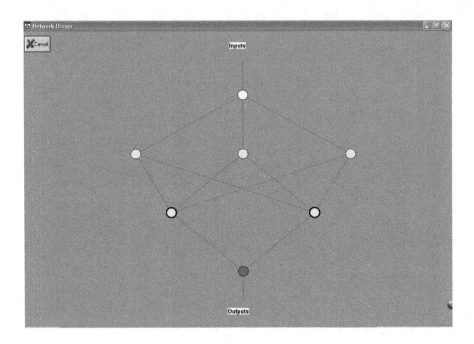

Figure A-16

Amplifier Neural Network Model

The Bayesian network shown in figure A-17 was constructed using the test data, network structure of figure A-15 and neural network of figure A-16. The network was constructed using a manual bin setting of 40 bins, normal distribution and average variance.

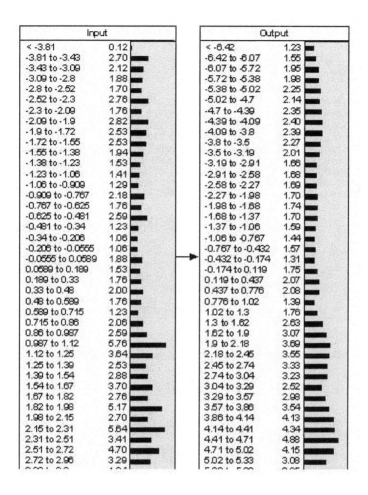

Figure A-17

Amplifier Cycle Computer-Constructed Network

The Bayesian network shown in figure A-18 was constructed using the step test data, network structure of figure A-15 and neural network of figure A-16. The step network used the average test database of 21 points. The network was constructed using a selection of 20 bins, normal distribution and nearest value variance.

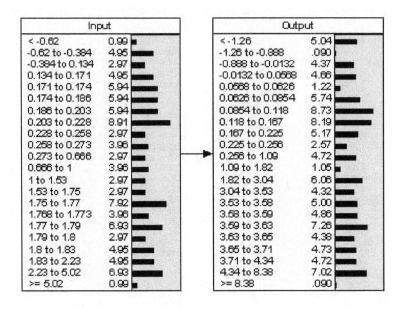

Input	
< -0.62	0.99
-0.62 to -0.384	4.95
-0.384 to 0.134	2.97
0.134 to 0.171	4.95
0.171 to 0.174	5.94
0.174 to 0.186	5.94
0.186 to 0.203	5.94
0.203 to 0.228	8.91
0.228 to 0.258	2.97
0.258 to 0.273	3.96
0.273 to 0.666	2.97
0.666 to 1	3.96
1 to 1.53	2.97
1.53 to 1.75	2.97
1.75 to 1.77	7.92
1.768 to 1.773	3.96
1.77 to 1.79	6.93
1.79 to 1.8	2.97
1.8 to 1.83	4.95
1.83 to 2.23	4.95
2.23 to 5.02	6.93
>= 5.02	0.99

Output	
< -1.26	5.04
-1.26 to -0.888	.090
-0.888 to -0.0132	4.37
-0.0132 to 0.0568	4.66
0.0568 to 0.0626	1.22
0.0626 to 0.0854	5.74
0.0854 to 0.118	8.73
0.118 to 0.167	8.19
0.167 to 0.225	5.17
0.225 to 0.256	2.57
0.256 to 1.09	4.72
1.09 to 1.82	1.05
1.82 to 3.04	6.06
3.04 to 3.53	4.32
3.53 to 3.58	5.00
3.58 to 3.59	4.86
3.59 to 3.63	7.26
3.63 to 3.65	4.38
3.65 to 3.71	4.73
3.71 to 4.34	4.72
4.34 to 8.38	7.02
>= 8.38	.090

Figure A-18

Amplifier Step Computer-Constructed Network

A.5 Model Comparison

The accuracy of each of the models for input values in figures A-9 and A-10 as compared to the test data is provided in figures A-19 and A-20 respectively. Data are compared for each test point at the maximum sampling rate of the test equipment which is 0.02 seconds.

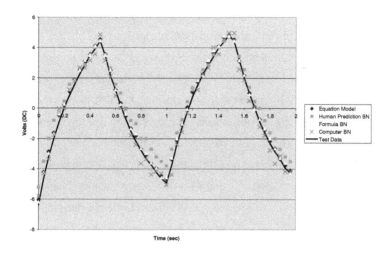

Figure A-19

Amplifier Cycle Model Comparison

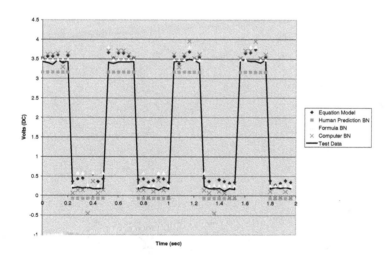

Figure A-20

Amplifier Step Model Comparison

134

As can be seen in figure A-19, the amplifier amplifies the cyclic wave at very nearly the gain of two with no observable non-linearity. All models do a good job of predicting this function. The step function shown in figure A-20 has a non-linear region at the lower portion of the step. This can be seen in figure A-10 as a small voltage of about 0.2 volts. As can be seen in figure A-20, this same 0.2 volt reading is also present at the output. The fact that there was no amplification as the signal was fed through the amplifier supports a conclusion that this is noise at the input and output of the amplifier. The formula models both amplified this input resulting in predictions that were high at the lower portion of the step. The equation-based Extend model and the formula BN had nearly identical results for the cyclic waveform but were slightly different for the step waveform. This was due to the discretization of the continuous data in the formula BN. The human judgment and computer constructed networks both learned this phenomenon from the learning data and did not amplify the noise. The human judgment network underestimated the voltage at both the high and the low step.

The average percent error for each model is presented in figure A-21.

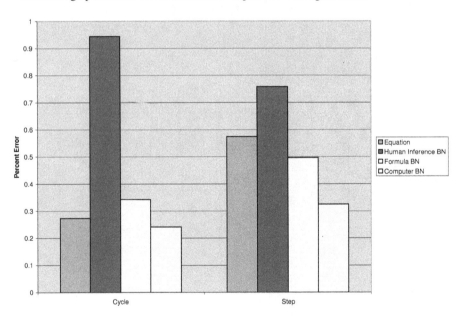

Figure A-21

Amplifier Average Percent Error

135

The computer generated Bayesian network had the lowest error while the human judgment Bayesian network had the highest error in both cases.

The time to complete the tasks required to build each model was recorded to the nearest minute. The results are presented in table A-1. The formula Bayesian network had the lowest construction time while the human inference network had the highest.

Table A-1

Amplifier Model Construction Times

Action in minutes	Equation-Based Model	Human Inference Bayesian Network	Formula Bayesian Network	Machine Learning Bayesian Network
Construct Model	8	5	5	1
Prepare Data Survey	-	11	-	-
Collect Data	-	83	-	-
Process Survey Data	-	36	-	-
Construct Neural Network	-	-	-	7
Computer Generation of Final Model	-	-	-	12
Update Probabilities	-	3	-	-
Total Time (mins)	8	138	5	20

As can be seen from table A-1, the collection and processing of data from human experts is a time consuming process. On average, it took approximately 8 minutes per person to make the predictions and 4 minutes to process the data.

Appendix B - LRC Circuit Models

The models of this appendix are a virtual representation of an electrical circuit with an inductor, capacitor and resistor wired together in series and driven by a 9-volt battery. The electrical wiring schematic of the circuit is provided in figure B-1. The objective of the models is to predict the voltage across resistor R when the switch is closed at time zero.

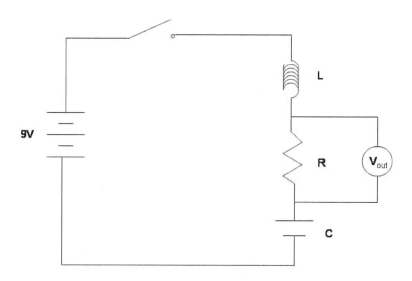

Figure B-1

LRC Circuit Schematic

The inductor, sometimes called a choke, resists a change in current flow. This results in a delay in the rise of the voltage when the switch is closed. The resistor determines the current flow in the circuit. The capacitor is a voltage storage device. If the capacitor has no charge, the full voltage will be present across the capacitor. As the capacitor charges, the voltage will decrease until it approaches zero when fully charged.

B.1 Equation-Based Model

The equation for this circuit is

$$Z(s) = R + s*L + 1/(s*C)$$

Solving the equation using initial condition $Z(0) = \infty$ yields

$$s^2*L + s*R + 1/C = 0.$$

Using the quadratic equation and solving for the roots yields

$$s = -R/(2*L) \pm [R^2/(4*L^2) - 1/(L*C)]^{1/2}.$$

This equation has three solutions. If the two roots are real and distinct, the solution is

$$V = A_1/R * \exp(s_1*t) + A_2/R * \exp(s_2*t)$$

$$A_2 = -A_1$$

$$A_1 = [V_0/(L*s_1)]/[1-(s_2/s_1)].$$

The other two solutions are for roots real and equal and for roots that are complex. For the purpose of this model, only real and unequal root solutions are considered. The Extend equation model of this circuit is provided in figure B-2.

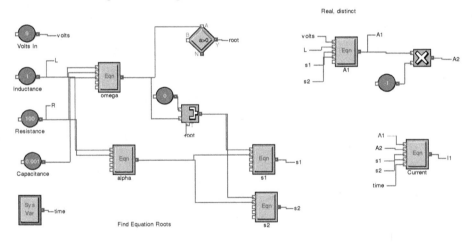

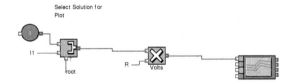

Figure B-2

LRC Equation Model

139

In this case, there are no model elements for the components of the circuit. The values of voltage (V), inductance (L), resistance (R) and capacitance (C) are individual inputs which are fed through a series of equation blocks. Roots s1 and s2 are determined and the real solution for A1 and A2 is solved. These values are used, along with time, to solve for the current in the circuit. The voltage is then determined from the current and resistance.

B.2 Human Judgment Bayesian Network

The manually constructed structure of the Bayesian network model for human judgment is provided in figure B-3.

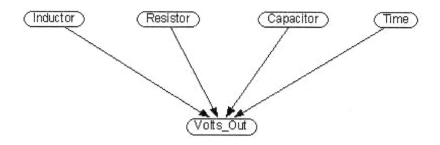

Figure B-3

LRC Human Judgment Bayesian Network Model

The probability tables were created through multiple predictions from seven volunteers who were provided with the test data shown in figure B-4. Test data were sampled using a Lego® RCX computer coupled with a high accuracy voltage measuring interface from Lego Dacta®. Data were captured using LabVIEW® Software from National Instruments. The number of samples was determined by using the Monte Carlo sampling technique described in section 2.1 for a width of a 0.1 volts and 95% confidence. The inductors, capacitors and resistors were purchased in standard, mixed lots of component values. A two second measurement period was selected for test measurements. Using the derived equations, a subset of components was selected that provided responses that could be measured over a two second period. The components for the tests were then randomly selected from this subset. A deliberate attempt was made to find an inductor value that would provide a measurable delay in the voltage rise. Unfortunately, in this application a value of approximately one Henry would be required while the largest commercial value was 300 micro Henries. The minimum sampling rate of 0.02 seconds was too great to measure the delay. This resulted in test data where changing the value of the inductor had no measurable impact on the output of the circuit.

Eight combinations of input variables were tested with the results provided in figure B-4.

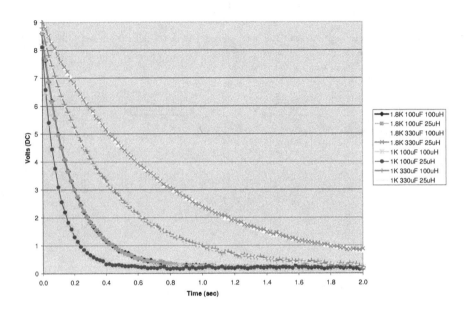

Figure B-4

LRC Test Data

These data were provided to the test volunteers. The volunteers were asked to provide predictions for the following different circuit values: 1KΩ 220 μF 100 μH, 2.8KΩ 220 μF 100 μH, and 2.8KΩ 220 μF 300 μH. The three cases were chosen to look at predictive accuracy where one, two and all three input values were different than those shown to the participants. This provided three problems that became gradually more difficult to predict.

The volunteers provided the predictions shown in figures B-5 through B-7. Test data, which the volunteers did not see, are provided for comparison.

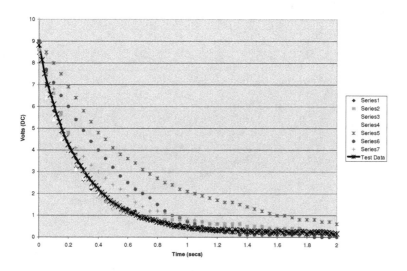

Figure B-5

1KΩ 220 μF 100 μH Predictions

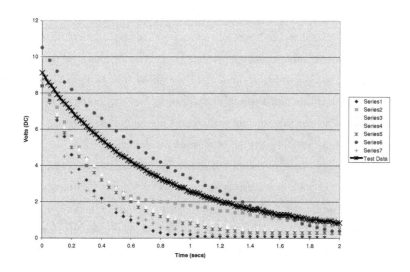

Figure B-6

2.8KΩ 220 μF 100 μH Predictions

142

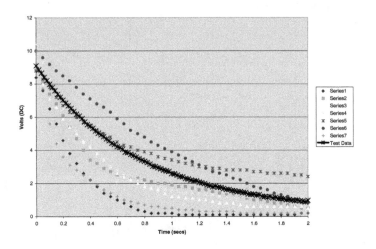

Figure B-7

2.8KΩ 220 μF 300 μH Predictions

Participants were also asked to provide a confidence assessment for their prediction. Assessments were on a scale of 1 to 5 where 5 corresponds to very high confidence and one corresponds to very little confidence in the prediction. This confidence was then multiplied by one hundred and used as the number of times that the prediction was counted during learning of probability tables. This resulted in a non-uniform prior for learning of the probability tables, ensuring that errors measured from this network were the result of the construction method and not residual probabilities caused by having only seven predictions.

The confidence assessments had mixed results when compared to predictive accuracy. The 1KΩ 220 μF 100 μH predictions results were correlated between confidence and predictive accuracy as shown in figure B-8. All but one confidence assessment in the 2.8KΩ 220 μF 100 μH predictions was ranked as three. The 2.8KΩ 220 μF 300 μH predictions showed an inverse relation between confidence and predictive accuracy as shown in figure B-9.

143

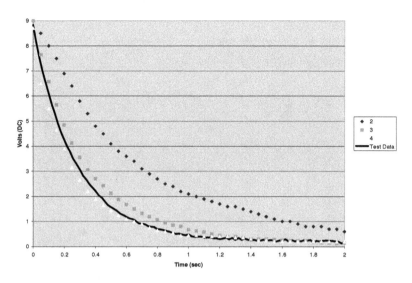

Figure B-8

LRC 1KΩ 220 μF 100 μH Confidence versus Accuracy

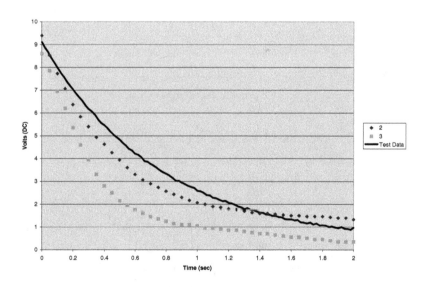

Figure B-9

LRC 2.8KΩ 220 μF 300 μH Confidence versus Accuracy

144

B.3 Formula Bayesian Network

Another Bayesian network model was created using the same formulae as those used for the equation based model. The formula Bayesian network is presented in figure B-10.

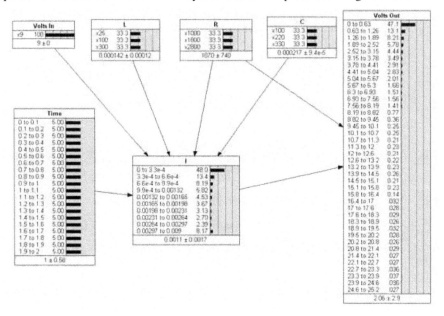

Figure B-10

LRC Formula Bayesian Network

The formula Bayesian network model was first constructed using intermediate variables like the equation model of figure B-2. When the equation to table option was executed, the variable "Volts Out" contained a mixture of real and imaginary numbers. This was caused by a second order equation with both real and imaginary roots. When the equation to table option is executed, each node with a formula calculates outputs for every combination of inputs. Some combinations resulted in imaginary numbers which continued through calculations to the "Volts Out" node. The network was modified to put all formulae into node "I" of figure B-10 to avoid the generation of imaginary numbers. This is a limitation of the Netica software package.

145

B.4 Computer-Constructed Bayesian Network

The final model created using the software described in section 6.1 is shown in figure B-11. The network structure was determined using the BN PowerConstructor program. The program correctly found that the value of the inductance (L) had no effect on the outcome.

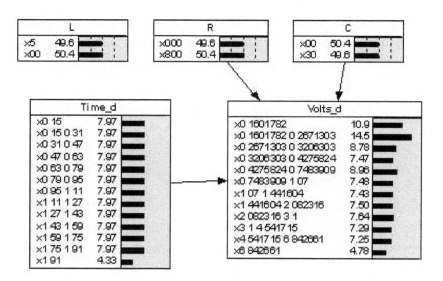

Figure B-11

LRC BN PowerConstructor Model

The BN Builder program required a trained neural network as an input to predict cases that were not contained in the input training set. These cases were the same ones the human volunteers were asked to predict. The input neural network was a fully-connected hybrid network containing two hidden layers with a sigmoid transfer function in the first and output layers and a Gaussian transfer function in the second layer. The network structure is presented in figure B-12.

146

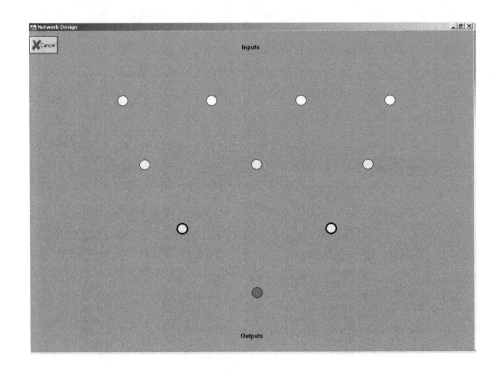

Figure B-12

LRC Neural Network Model

The network was trained using the average data shown in figure B-4. This consisted of a data set with 808 observations.

The Bayesian network shown in figure B-13 was constructed using the test data, network structure of figure B-11 and neural network of figure B-12. The network used the entire test database of 11339 observations. The network was constructed using a fixed number of 40 bins, normal probabilities and average variance as input options.

147

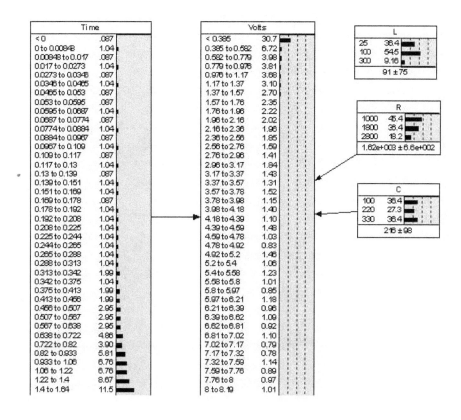

Figure B-13

LRC Computer-Constructed Network

B.5 Model Comparison

The accuracy of each of the models compared to the test data is provided in figures B-14 through B-16.

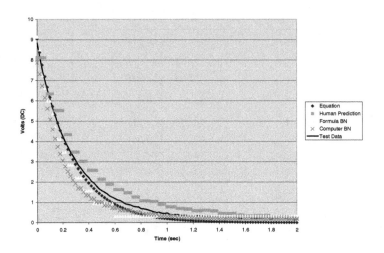

Figure B-14

1KΩ 220 μF 100 μH Model Comparison

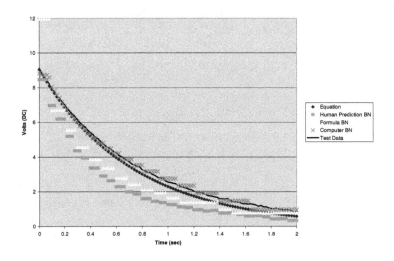

Figure B-15

2.8KΩ 220 μF 100 μH Model Comparison

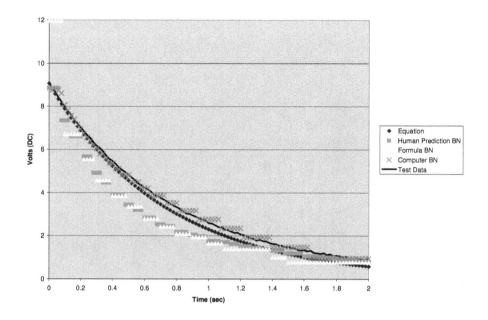

Figure B-16

2.8KΩ 220 µF 300 µH Model Comparison

The average percent error from the mean test data for each model is presented in figure B-17.

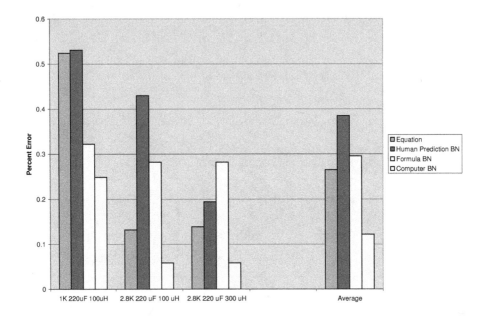

Figure B-17

LRC Model Percent Error Comparison

On average, the human judgment model had the worst accuracy while the computer generated Bayesian network had the best accuracy.

The time to complete the tasks required to build each model was recorded to the nearest minute. The results are presented in table B-1.

Table B-1

LRC Model Construction Times

Action (minutes)	Equation-Based Model	Human judgment Bayesian Network	Formula Bayesian Network	Machine Learning Bayesian Network
Develop Equations	27	-	27	-
Construct Model	66	4	76	3
Prepare Data Survey	-	47	-	-
Collect Data	-	85	-	-
Prepare Survey Data	-	183	-	-
Construct Neural Network	-	-	-	5
Computer Generation of Final Model	-	-	-	41
Update Probabilities	-	13	-	-
Total Time	93	332	103	49

The human judgment model had the highest construction time while the computer generated model had the lowest. The average time per human prediction was approximately 10 minutes while the average to process the data was 22.

Appendix C - NACA Wing Models

This set of models is a virtual representation of the lift coefficient produced by wing airfoil shapes. The National Advisory Committee for Aeronautics (NACA), the forerunner of NASA, created a four digit nomenclature to identify wing shapes. All numbers in the four-digit NACA wing identifiers are expressed in relation to the length of the wing chord which is the distance from the tip of the leading edge to the trailing edge of the wing. The first digit identifies the maximum height of the wing mean line, a line equidistant between the upper and lower wing surface, expressed as a percentage of the chord. The second digit identifies the distance from the leading edge that the maximum height is located expressed in tenths of the chord. The third and fourth digits identify the maximum thickness of the airfoil expressed as a percentage of the chord. These numbers are illustrated in figure C-1

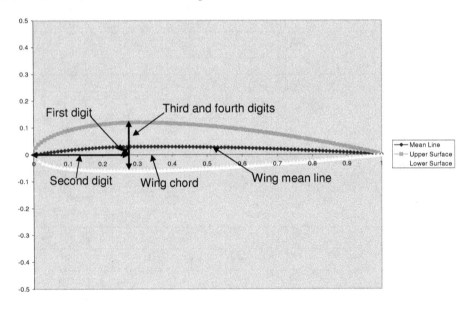

Figure C-1

NACA 2318 Wing Example

A NACA-2318 wing has the mean line a maximum of 2% above chord with this maximum located 30% from the leading edge. The maximum wing thickness is 18% of the chord.

C.1 Equation-Based Model

The equation model for predicting the lift coefficient was created by Kevin Jones of the Aeronautical Engineering Department at the Naval Postgraduate School. This model is a public domain model and is used with permission from the author. It can be viewed at http://www.aa.nps.navy.mil/~jones/online_tools/panel2/. The model uses a panel code method to compute the pressure differential over the wing surface by partial solution of the Navier-Stokes Equations for fluid flow. Direct solution of the entire system of equations is not feasible, even with today's powerful computers. The coefficient of lift (C_L) is then calculated from this pressure data. The panel code method is limited to angles-of-attack in which flow separation does not occur and subsonic airspeeds. Further information on the formulae and methods used can be found at the above web address. The air pressure distribution for a NACA 2318 wing at 6.0 degrees angle-of-attack (AOA) is shown in figure C-2. The angle-of-attack is the angle between the chord of the wing and the air stream impinging on the leading edge. All data collected from this model were collected at a resolution of 100 panels, the highest possible resolution.

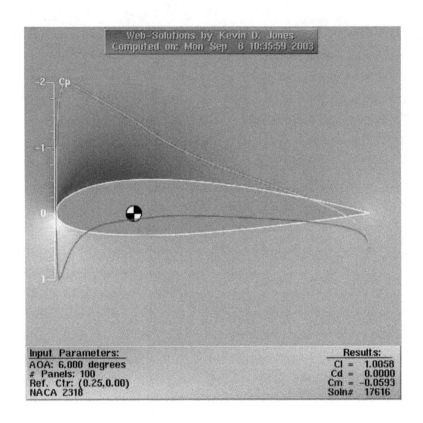

Figure C-2

NACA 2318 Air Pressure Distribution

C.2 Human Judgment Bayesian Network

The manually constructed Bayesian network structure for human judgment is provided in figure C-3.

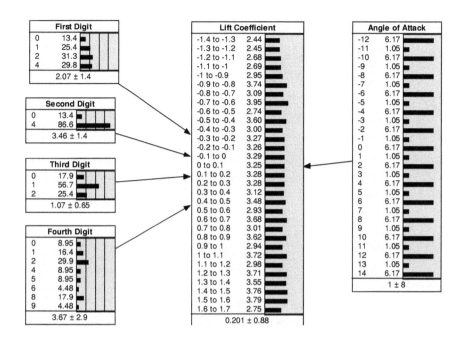

Figure C-3

Human Judgment Bayesian Network Model

The probability tables were created through multiple predictions from eight volunteers who were provided with the test data shown in figures C-4 and C-5. Test data were taken from the book The Theory of Wing Sections by Abbott and Doenhoff and are based on wind tunnel tests. All test data used is measured at a Reynolds Number of 6.0×10^6 which provides a reasonable match to the limitations of the equation model. No accuracy was provided for the measurements.

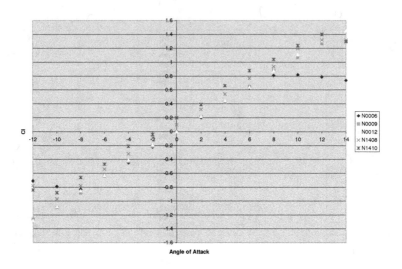

Figure C-4

NACA Thin Wing Wind Tunnel Test Data

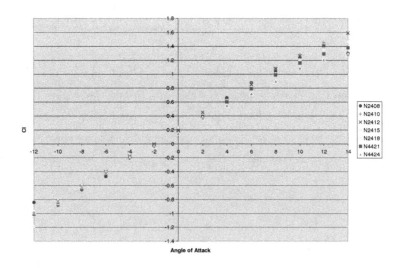

Figure C-5

NACA Thick Wing Wind Tunnel Test Data

157

The volunteers were asked to provide predictions for NACA 1412 and NACA 4421 wings. These two wings were chosen so that one wing is a relatively thin airfoil while the second was a relatively thick airfoil. The exact wing in each category was chosen at random. The volunteers provided the predictions shown in figures C-6 and C-7. Test data, which the volunteers did not see, are provided for comparison.

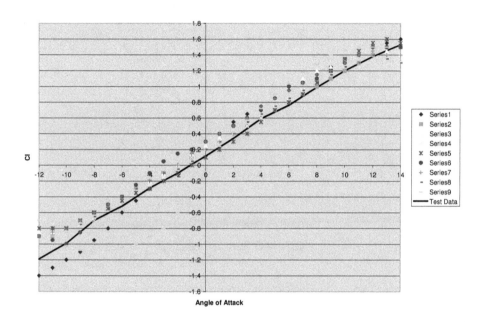

Figure C-6

NACA 1412 Human Judgment Predictions

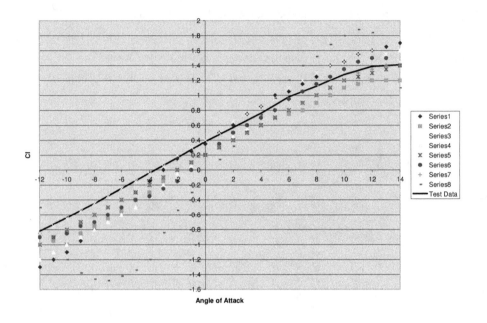

Figure C-7

NACA 4421 Human Predictions

Participants were also asked to provide a confidence assessment between one and five for their prediction with five assessing high confidence in their prediction and one assessing low confidence in the prediction. Each confidence assessment was multiplied by 100 and to determine how many times that prediction was counted during learning of the probability tables. The confidence assessments again did not have a strong correlation with predictive accuracy. The comparisons for the NACA 1412 and 4421 wings are shown in figure C-8 and C-9.

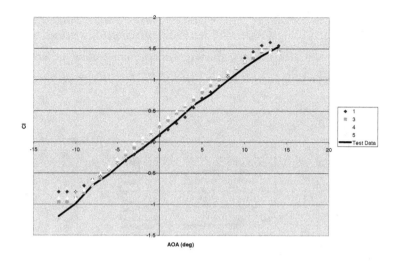

Figure C-8

NACA 1412 Confidence versus Accuracy

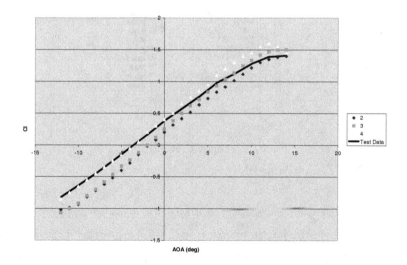

Figure C-9

NACA 4421 Confidence versus Accuracy

C.3 Computer-Generated Bayesian Network

A computer model using the software described in section 6.1 was used to create a computer-generated model. The network structure was determined using the BN PowerConstructor program and is presented in figure C-10.

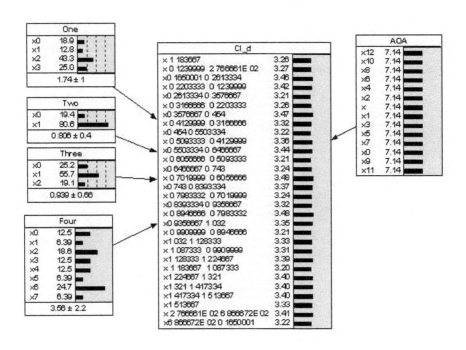

Figure C-10

Wing BN PowerConstructor Model

A trained neural network was required as an input to predict the two test cases. These two cases were the same wing shapes the human volunteers were asked to predict. The neural network was a fully-connected hybrid network containing two hidden layers with a sigmoid transfer function in the first and output layers and a Gaussian transfer function in the second layer. The network structure is presented in figure C-11.

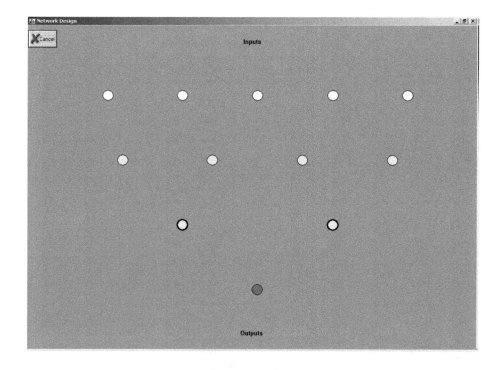

Figure C-11

Wing Neural Network Model

The network was trained using the data shown in figures C-4 and C-5. The training data consisted of 252 observations.

The computer constructed Bayesian network shown in figure C-12 was constructed using the test data, network structure of figure C-10 and neural network of figure C-11. A fixed number of 40 bins were used and probabilities were created using the Netica learning algorithm. The normal probability option was not programmed for a single continuous node with all discrete parents. The probability tables were updated using the output learning set of the program and a frequency of 5000; the same weight if ten volunteers had all assessed a confidence of 5. This number insured that accuracy comparisons were not influenced by residual probabilities in node "CI" caused by the high number of bins selected.

162

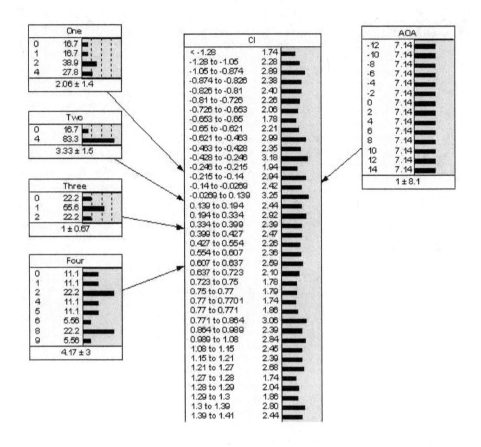

Figure C-12

Wing Computer-Constructed Network

C.4 Model Comparison

The accuracy of each of the models compared to the test data is provided in figures C-13 and C-14.

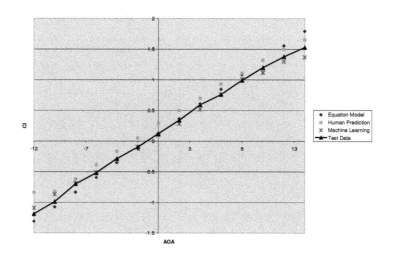

Figure C-13

NACA 1412 Model Comparison

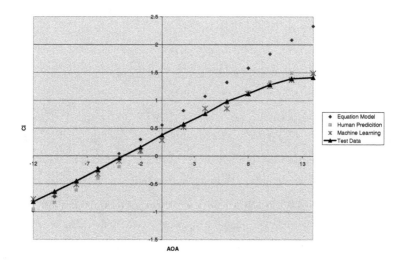

Figure C-14

NACA 4421 Model Comparison

As can be seen in figure C-14, flow separation begins to occur at about six degrees AOA with an associated loss of lift. The average percent error for each model is presented in figure C-15.

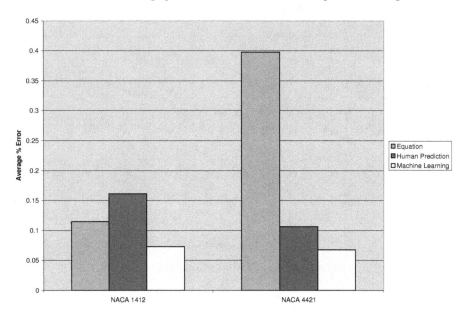

Figure C-15

Model Percent Error Comparison

All three models had similar accuracy for the thinner NACA 1412 wing. The equation-based model had much higher error for the NACA 4421 wing. This was caused by flow separation that caused a loss of lift at extremely low or high angles of attack. The equation model could not predict that this loss of lift would occur. Both the human judgment predictions and the computer-generated network learned that this phenomenon would occur in thicker wings, resulting in a more accurate prediction than the equation-based model. The computer constructed Bayesian network had the lowest error in both cases.

The time to complete the tasks required to build each model was recorded to the nearest minute. The results are presented in table C-1.

Table C-1

Wing Model Construction Times

Action (minutes)	Equation-Based Model	Human judgment Bayesian Network	Machine Learning Bayesian Network
Develop Equations	-	-	-
Construct Model	480*	4	3
Prepare Data Survey	-	47	-
Collect Data	-	77	-
Process Survey Data	-	183	-
Construct Neural Network	-	-	5
Computer Generation of Final Model	-	-	41
Update Probabilities	-	13	1
Total Time	480*	324	50

* The time for the equation model is an estimate from the author to build the model today using current modeling tools.

Note that in these models the human judgment still has a high time associated with collecting the human predictions (307 minutes), but the total time is still less than that estimated to create the equation-based model. The average time to make the two predictions was 11 minutes per person and the average time to process one pair of predictions was 26 minutes. The data survey preparation time of 47 minutes is not effected by the number of predictions. The time to collect a single pair of predictions would be approximately 84 minutes.

Appendix D - Elevator Control Models

This set of models is a logical representation of the control system for an elevator. The model is constructed to demonstrate the differences in construction time between rules programmed in an equation-based modeling and simulation package and the same rules implemented in a Bayesian network. Because all models implement a fixed set of rules, no optimization is required. Accuracy is measured as the outcome of movement to determine if the elevator moved to the correct floor. For the computer constructed network, it is assumed that a measuring device was placed on the elevator to record its operation over a time period sufficient to capture 1612 elevator movements. Because attaching a recording device to an elevator control could possibly potentially impair its operation causing a safety issue, the data for the computer-generated Bayesian network was generated by recording the elevator movements of the equation-based model implemented in rules.

There are four levels (ground, one, two and three) in the building that the elevator services. The elevator has four selection buttons, one for each floor. The ground and third floor each has a single call button. The first and second floors each has two call buttons; one for up and one for down. Each button can be either selected or not selected. The elevator can be headed in either the up or down direction of travel. It will be at one of the four floors when a decision must be made as to which floor to move to next. If no buttons are selected, the elevator remains at its current position. There are therefore 8192 possible combinations of these variables. The control logic of the elevator is to continue to move in a single direction stopping where either a floor has been selected or a call button has been activated requesting movement in the direction the elevator is traveling. The elevator stops at the closest floor moving in its current direction. When the elevator has either reached the ground or third floor or there are no buttons selected in the current direction, the elevator will reverse direction and begin the same logic in the opposite direction.

D.1 Equation-Based Model

The first part of the equation model, shown in figure D-1, determines whether a button has been pushed and which floors are currently requesting a stop.

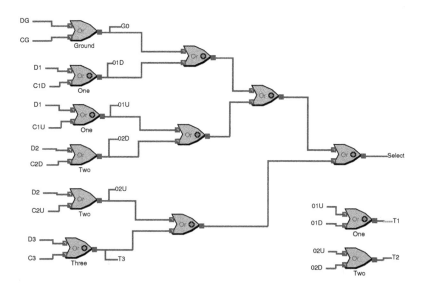

Figure D-1

Elevator Button Selection

The buttons with a "D" prefix represent a selected destination while the "C" prefix represents a call button. The suffixes "U" and "D" indicate the up or down directions. The output of the left column of OR gates is a generic set of conditions used to develop the rules. For example, in the upper left block the output "G0" comes from input conditions ground destination button selected or ground call button selected. The same rule will apply if either or both of these conditions are present. This allows a set of 512 rules to be developed to cover the 8192 possible combinations of conditions.

The next portion of the equation model determines the direction and current position of the elevator and routes the program to the correct set of rules for that position. This is presented in figure D-2.

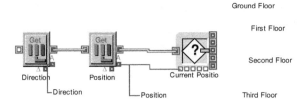

Figure D-2

Elevator Position Rule Selection

Figure D-3 displays the rule model for the ground floor. There is a similar model for each of the other three floors. The first block prevents the elevator from moving if either ground floor is selected or the ground floor call button is active since the elevator is already on the ground floor. The second block sets the direction to "up" since that is the only direction the elevator can travel. The third block checks to see whether any button has been depressed. If no button is depressed, the elevator does not move and the control loops back to the beginning. If a button has been depressed, the program is routed to the fourth block to determine which button(s) are depressed. The fourth block determines to which floor the elevator will move. This block starts from the top down and moves to the floor of the highest priority button selected. The remaining blocks choose the floor and direction. For example, if the first floor down call button were selected, the program would check first to see if either floors 1, 2, or 3 or call buttons 1 up, 2up or 3 were selected and then take the fourth branch which selects destination first floor and resets the direction to "down".

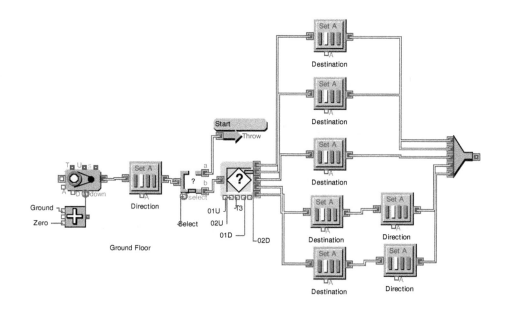

Figure D-3

Elevator Move Rules

D.2 Manually Constructed Bayesian Network

The manually constructed Bayesian network model is provided in figure D-4.

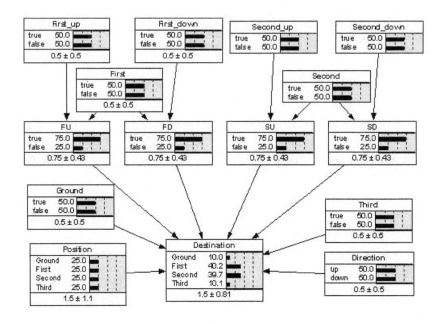

Figure D-4

Elevator Manual Bayesian Network Model

The probability tables were created by manually inserting the values into the network CPT tables by the author. This was accomplished by manually entering a probability of 100% in each row of the probability table for node "Destination" for the correct floor based on the input values. A zero was entered into all other cells of the row. As can be seen in figure D-4, the nodes "FU", "FD", "SU" and "SD" serve the same function as the OR gates in the equation model reducing the total number of rules required to cover all conditions. All other nodes were set to uniform probability.

D.3 Computer-Generated Bayesian Network

A model constructed by machine learning is presented in figure D-5. The relationship was constructed using the BN PowerConstructor program.

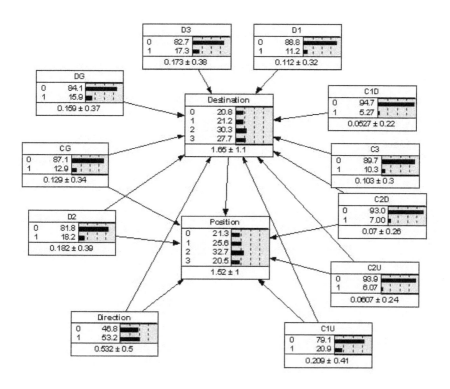

Figure D-5

Elevator BN PowerConstructor Model

The network was trained using data generated by the equation model output. This consisted of 1612 data sets. This data set consisted of all discrete values and had no missing values. Use of the BN Constructer program was not required.

The accuracy was the same for all models with each model sending the elevator to the correct floor. The time to complete the tasks required to build each model was recorded to the nearest minute. Times are for construction of the control system only and do not include integration with the elevator model for final test. The results are presented in table D-1.

Table D-1

Elevator Control Construction Times

Action (minutes)	Equation Model	Human Inference Bayesian Network	Machine Learning Bayesian Network
Develop Rules	9	9	9*
Construct Model	197	120	8
Total Time	206	129	17

* Because the data was generated from the simulation instead of recording it, the time used to manually record the rules is assessed to the computer-generated model.

E. Radar Models

This set of models is a virtual representation of an airborne radar. Radar is a search and tracking system that detects targets via electromagnetic transmission and reflection. Pulses of radio energy are emitted by the radar transmitter into the air. If a radar reflective target is within the range of the radar, the pulse is reflected back from the target and is picked up by the radar receiver. For airborne radar, the transmitter and receiver are collocated. There are a number of variables that affect the maximum range a specific radar system can detect a target. The AN/APG-66(V)2 radar, used on foreign export model F-16 aircraft, is the system that is modeled for this problem. All data are unclassified.

Detecting a target by radar is a very complex phenomenon. There are two general properties of the influences that affect the radar signal: gains and losses. Gains strengthen the signal and increase the range of the radar. An example of a gain is the use of a focused antenna that concentrates the signal into a smaller area of space. Losses weaken the signal decreasing the range of radar. Covering the antenna with a radome to protect the radar and provide an aerodynamic outer surface to the fuselage of the aircraft results in losses as the signal must pass through the covering.

E.1 Equation-Based Model

To predict the maximum range of specific radar, the radar range equation has been developed. This equation does not capture the entire complex phenomenon involved in actual radar operation. It is a useful design tool to predict the relative magnitude of change a system will experience when a parameter in the equation is changed. It is an excellent example of a system for which complete equations to describe its operation currently do not exist. The radar range equation[7] is

$$R_{max} = [(P * G^2 * \sigma * \lambda^2) / ((4\pi)^3 * L * P_{min})]^{1/4}$$

R_{max} is the maximum range

P is the transmitter power

G is the antenna gain

σ is the target radar cross section

λ is the radar wavelength

---

[7] Equation from [Cohen et al. , 1992].

L is the loss coefficient

P_{min} is the minimum receiver power

Based on the published information in the Pilot's Manual for the AN/APG-66(V)2 Fire Control Radar as Installed In the U.S. Naval Test Pilot School's Airborne System Training and Research Support Aircraft manual, the factors affecting the losses are

$$L = KT0 * Fn * Bn * RL * Squint * Xmit/Rec * Atmos * Fspace$$

KT0 is Boltzmann's constant times thermal noise

Fn is the noise figure for the radar

Bn is bandwidth noise

RL is radome losses

Xmit/Rec is transmitter/receiver losses

Atmos is atmospheric losses

Fspace is free space losses

The radar range equation applies to a spotlight radar that continuously shines its beam on the target with the target in the center of the beam. The actual radar tests were conducted with the radar in search mode. The beam of the radar in search mode is constantly moving over a fixed search volume in search of the target. Thus, the target will be illuminated for only a small portion of each search. As a result, the radar range equation would provide an over optimistic prediction for a radar that is scanning a search volume.

In order to better match the equation to a radar in search mode, a factor for search volume which approximates the effects of spreading the radar power over the search volume. This introduces the terms θ_a / θ_{bw} and θ_e / θ_{bh} where θ_a is the horizontal sweep angle, θ_{bw} is the radar beam width, θ_e is the vertical sweep angle and θ_{bh} is the beam height [Stimson, 1983]. This modified version of the equation has the effect of spreading the radar power over the scan volume for each search scan. Although not exact, this provides some additional losses to the denominator of the radar range equation to account for the scanning motion. All data were collected in one bar 10 degree search mode. This equates to an elevation sweep angle equal to the beam height and a sweep angle of 20 degrees ($\pm$10 degrees of the nose). The beam width is 3.25 degrees.

Since the final equation is complex and contains both very large and very small numbers, a more convenient solution is to implement the equation in decibels. A decibel is defined as

175

$$dB = 10 * \log_{10}(x)$$

where x is any of the variables in the radar range equation. The numerator terms which are the gains of the system for the equation model of the radar are shown in figure E-1.

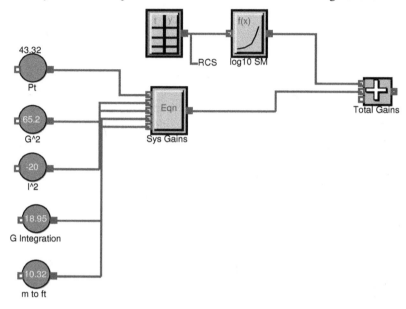

Figure E-1

Radar Equation Model System Gains

The denominator terms which are the losses in the system are shown in figure E-2.

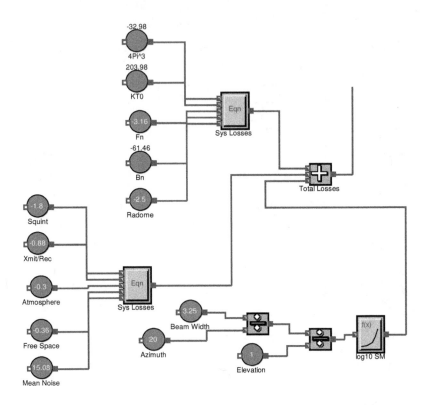

Figure E-2

Radar Equation Model System Losses

The gains and losses are combined and the output is converted from dB to nautical miles as is shown in figure E-3.

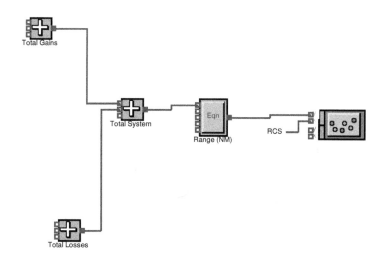

Figure E-3

Radar Equation Model Conversion to Nautical Miles

E.2 Formula Bayesian Network

Predicting radar range given input parameters would be a difficult problem for anyone but a radar expert. Since none was available, a manual Bayesian network model was created using the same formulae as the equation model. The manually constructed Bayesian network model is provided in figure E-4.

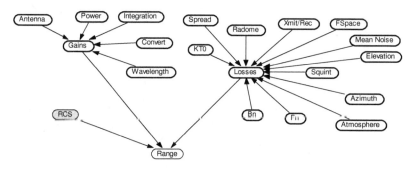

Figure E-4

Manual Bayesian Network Radar Model

178

The probability tables were filled in using the equation to CPT table function using the same radar equations as the model of figure E-1.

E.3 Computer-Generated Bayesian Network

For the computer-generated model, radar performance data for radars operating at similar frequencies were used and is presented in table E-1. The available data only covered a target up to one dBsm in size. Therefore, data for 10 and 20 dBsm were predictions obtained using the equation

$$R_2 = (R_1{}^4 * (\sigma_2 / \sigma_1))^{1/4}$$

where R_1 is the range at radar cross section σ_1 and R_2 is the calculated range at new σ_2. The neural network used in construction of the final model could not be used for these predictions because the required prediction values were well above the upper bounds of the data set. A neural network for the RCS and Range data was created and is presented in figure E-5. The neural network is a fully-connected hybrid network containing two hidden layers with a sigmoid transfer function in the first and output layers and a Gaussian transfer function in the second layer.

Table E-1[8]

Radar System Detection Ranges

Name	Freq (GHz)	RCS (dBsm)	Range (NMI)
RASIT-3190B	9	0.001	5
RASIT-3190B	9	0.01	9
RASIT-3190B	9	0.1	16
RASIT-3190B	9	1	28
RASIT-3190B	9	10	50
RASIT-3190B	9	20	89
Unknown	9.1	0.001	6
Unknown	9.1	0.01	10
Unknown	9.1	0.1	17
Unknown	9.1	1	31
Unknown	9.1	10	55
Unknown	9.1	20	98
AN/GPN-22	9.1	0.001	7
AN/GPN-22	9.1	0.01	12
AN/GPN-22	9.1	0.1	21
AN/GPN-22	9.1	1	31
AN/GPN-22	9.1	10	55
AN/GPN-22	9.1	20	98
PAR	10	0.001	5
PAR	10	0.01	9
PAR	10	0.1	16
PAR	10	1	28
PAR	10	10	50
PAR	10	20	89

---

[8] From [Kincaid, 1993].

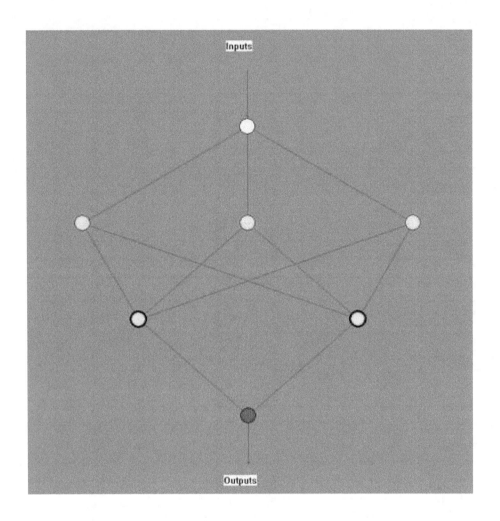

Figure E-5

Radar Neural Network

The computer-generated model created from the data of table E-1 and neural network of figure E-5 is shown in figure E-6.

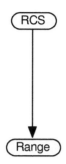

Figure E-6

Computer-Generated Bayesian Network Structure

This network demonstrates both an advantage and disadvantage of computer generated networks. In this case, it is not necessary to include all of the constants and intermediate variables of the radar range equation resulting in a much simpler model. The model can accurately predict the range of a target of a given radar cross section. However, changes in the output of a system caused by changing some of the variables not included in the Bayesian network model may be of particular interest in an engineering trade study.

A comparison of test data with model predictions is presented in figure E-7. Aircraft types are not identified as radar cross section measurements are classified.

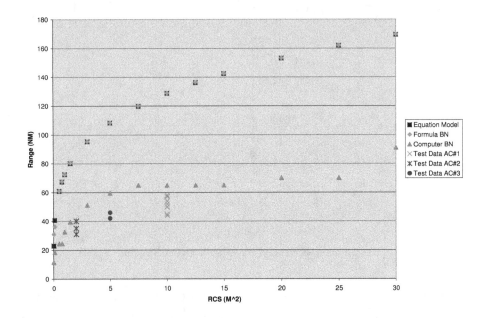

Figure E-7

Radar Model Accuracy Comparison

The average percent error for each model of figure E-7 is compared to the average detection range for each aircraft with results presented in figure E-8. As can be seen in figures E-7 and E-8, the radar range equation does a poor job of predicting actual radar range performance, even when the sweep correction is added.

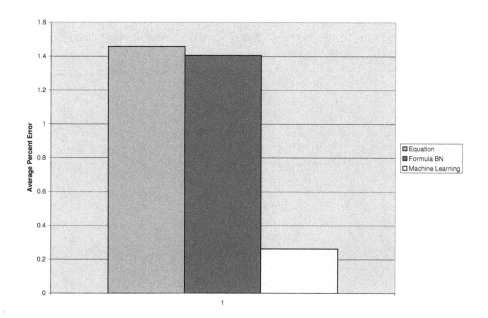

Figure E-8

Radar Model Average Error

The construction times for each model are presented in table E-2.

Table E-2

Radar Model Construction Times

Action (minutes)	Equation Model	Formula Bayesian Network	Machine Learning Bayesian Network
Develop Equations	124	124	-
Build Database	-	-	31
Construct Model	72	44	13
Total Time	196	168	44

The computer constructed Bayesian network had the lowest time of construction. The time for this particular model is higher than other computer generated networks due to the requirement to manually adjust the learning data set.

Appendix F - Forward Looking Infrared (FLIR) Models

This set of models is a virtual representation of the WF-360TL Forward Looking Infrared (FLIR) system. This is an unclassified commercial system installed on a P-3 aircraft. FLIR systems detect targets using heat (IR energy). The target must have a different temperature than the background in order for the target to be seen. The target may be either hotter or colder than the background. Because the FLIR detects IR energy, no visible light is required to detect an image. However, humidity in the air absorbs IR energy and degrades system performance. FLIR systems work best in dry air and when there is a very strong temperature differential between the target and background.

F.1 Equation-Based Model

FLIR systems are defined by two key specification parameters: the effective minimum resolvable effective temperature difference and the spatial cut-off frequency. The minimum resolvable temperature is the minimum difference between the temperature of an object and the temperature of the background behind the object for it to be visible on the display. The effective temperature accounts for the environmental effects of humidity. The greater the relative humidity, the greater the reduction in effective temperature difference from the actual temperature difference.

The spatial frequency (SF) is related to the size of the target. The SF formula is $SF = SR / (2000 * TW)$

$$SR = \text{slant range in feet}$$
$$TW = \text{target width in feet}$$

The spatial cut-off frequency SF_{co} is related to the field of view of the display. Based on the display size and resolution, there is a range beyond which a target will not be seen on the display no matter how hot it is because it is too small to be seen on the display. This FLIR unit has two display settings of Wide (WFOV) and Narrow (NFOV). The WFOV has a look area of 11.1 degrees vertically and 14.8 degrees horizontally and the NFOV area is 2.7 degrees vertically and 3.6 degrees horizontally. The display is a black and white video image. Hot objects may be displayed in black with cooler objects in white if "Black" is selected on the polarity setting. Similarly, hot objects are displayed as white with cooler objects in black if "White" is selected. For this system, the effective minimum temperature differential is 0.10 degrees C and the SF_{co} is

1.54 cycles/milliradian (mrad) in WFOV and 6.33 cycles/mrad in NFOV. The final point that defines the system is the minimum effective temperature at half the spatial cut-off frequency. The specification value for this system is 0.33 degrees C. The spatial frequency in cycles/mrad can be approximated by the formulae:

$$T1 = (1 / \cos(\pi / 4) - 1 / (\Delta T_{1/2} - \Delta T_{min})$$

$$T2 = \pi / 2 / SF_{co}$$

where T1 and T2 are temporary variables, $\Delta T_{1/2}$ is the minimum effective temperature difference at ½ the cutoff frequency, ΔT_{min} is the minimum effective temperature difference at the cutoff frequency and SF_{co} is the spatial cutoff frequency. The spatial frequency is approximated using

$$SF = (1 / T2) * a\cos(1 / (T1 * (E\Delta T - \Delta T_{min}) + 1))$$

where SF is the spatial frequency and $E\Delta T$ is the effective temperature difference between the target and background.

A chart of system performance for both field-of-view settings is provided in figure F-1.

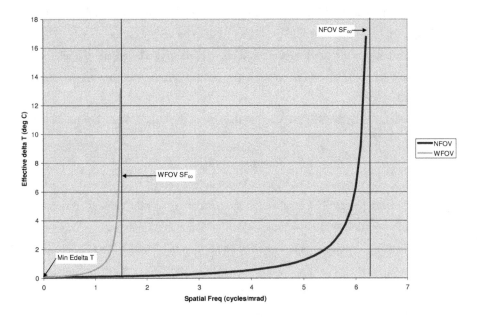

Figure F-1

WF-360TL FLIR System Performance

The final set of equations converts the spatial frequency to the range from the target. The maximum range in feet at which a target can be seen is

$$Range_{max} = SF_{max} * 2000 * TW$$

where SF_{max} = maximum spatial frequency for the $E\Delta T$ and TW = target width in feet.

Any target that is at a range less than the one calculated in the above equation should be visible on the display. The system has different spatial cutoff frequencies for the Wide and Narrow settings so each setting will have a different maximum range.

The equation model for predicting FLIR detection range is shown in figure F-2.

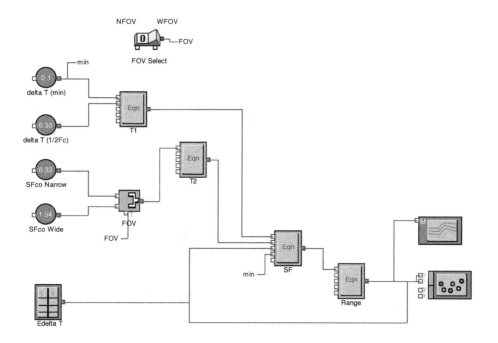

Figure F-2

FLIR Equation Model

The switch at the top allows selection of either WFOV or NFOV. The upper set of inputs provides the system specification values. The lower box "Edelta T" provides a range of effective temperature differential values to calculate the performance graphs on the right side. Maximum range performance of the equation model is shown in figure F-3.

FLIR Detection Ranges

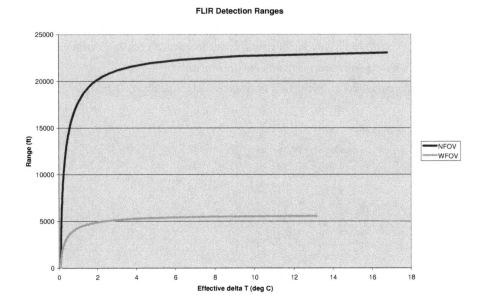

Figure F-3

FLIR Equation Model Maximum Range Predictions

There is a significant difference in maximum range of the system depending on which viewing mode is selected.

F.2 Formula Bayesian Network

A Bayesian network formula model was manually created using the same equations as the equation model. The network is shown in figure F-4.

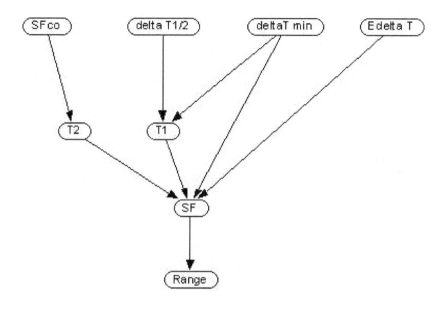

Figure F-4

Manual Bayesian Network FLIR Model

The input nodes were set at uniform probabilities and the other node probabilities were calculated using the equation to table option in Netica. The maximum range prediction for this model is presented in figure F-5. These predictions are very similar to those obtained with the equation model shown in figure F-3. Any differences are due to the discretization of the continuous data.

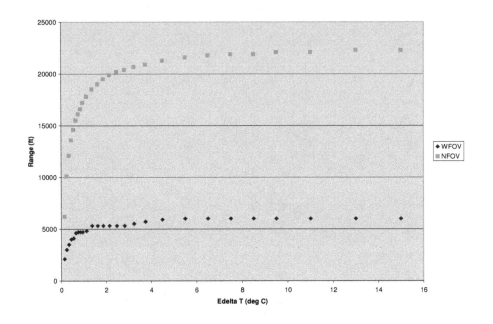

Figure F-5

FLIR Formula Bayesian Network Maximum Range Predictions

F.3 Computer-Generated Bayesian Network

Test data were provided by the U.S. Naval Academy Test Pilot School. Data were recorded by students during training flights and have a high degree of scatter. This is most likely due to measurement error as this was the first time many students had attempted to operate this equipment while recording the data. This lack of operator experience may also explain why the values recorded were consistently less than those predicted by the equations. FLIR displays require operator tuning of the system gain to optimize the display. The system also may not have been performing as well as possible because of age, coolant servicing or calibration issues. Data were measured by flying at a temperature controlled measurement board with alternating hot and cold bars 1.86 feet wide. When the operator could distinguish between two bars of different temperatures, the range and temperature delta were recorded. The effective temperature difference was calculated after correcting for altitude and relative humidity. Data were collected using both black hot and white hot polarity. Test data are displayed in figures F-6 and F-7.

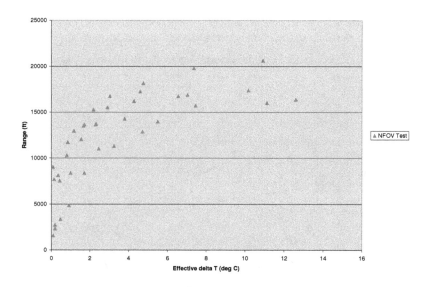

Figure F-6

FLIR NFOV Test Data

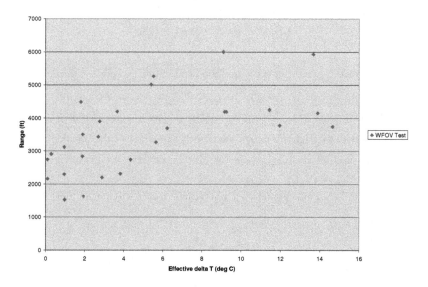

Figure F-7

FLIR WFOV Test Data

192

A computer-constructed Bayesian network model was constructed using the data of figures F-6 and F-7. The network structure determined by the BN PowerConstructor program is provided in figure F-8.

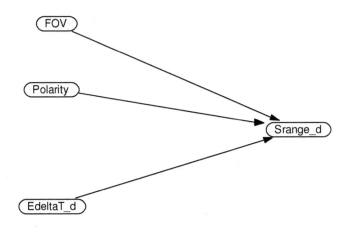

Figure F-8

FLIR BN PowerConstructor Model

Note that the BN PowerConstructor program found a relationship between the polarity of the display and the slant range of the system. Theoretically, there should be no difference in range caused by the polarity selection. This option is provided to allow a preference for the operator as to viewing objects with the higher temperature objects showing as either black or white on the display. However, plotting the data by polarity selection clearly shows two distinct sets of data as shown in figures F-9 and F-10.

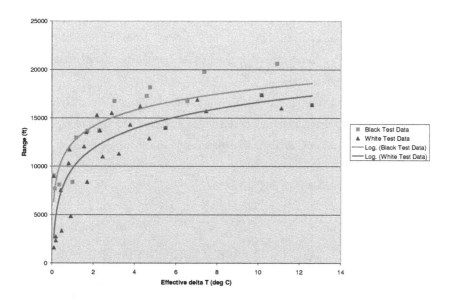

Figure F-9

FLIR Narrow FOV with Polarity

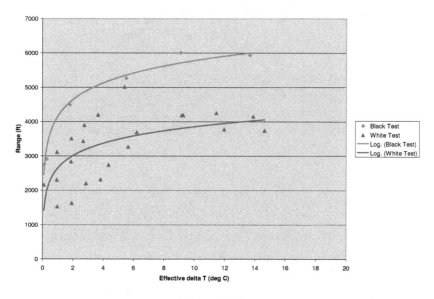

Figure F-10

FLIR Wide FOV with Polarity

194

Due to the unusual nature of the data, this discrepancy with theory was checked with the U.S. Navy Test Pilot School who confirmed the same observation. Since this anomaly does not appear in other systems, it is likely caused by a poorly designed display that generates a better picture in black hot than white hot.

The neural network for the FLIR model is presented in figure F-11. In this case, the neural network is used for data smoothing of the highly scattered data set and generation of additional points for the learning database to ensure good probability distributions.

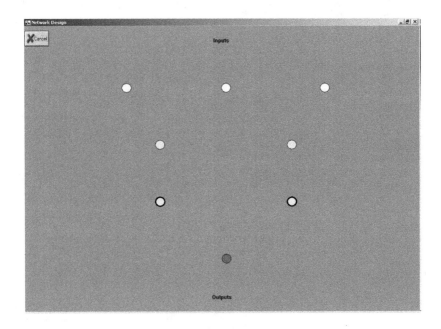

Figure F-11

FLIR Neural Network Model

The neural network is a fully-connected hybrid network containing two hidden layers with a sigmoid transfer function in the first and output layers and a Gaussian transfer function in the second layer. The final Bayesian network model, constructed using the data of figures F-6 and F-7, the structure of figure F-8 and the neural network of figure F-11 is presented in figure F-12.

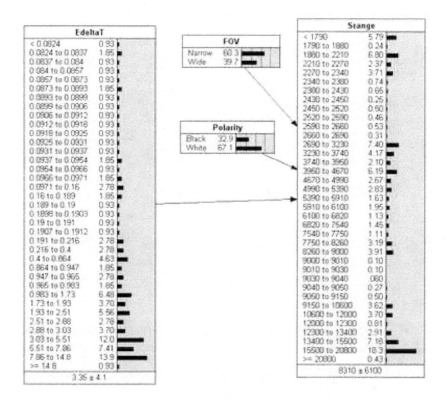

Figure F-12

FLIR Computer-Generated Bayesian Network Model

The network was trained using the test data consisting of 71 observations.

F.4 Comparison of Modeling Methods

Comparison of the models with the test data are provided in figures F-13 and F-14.

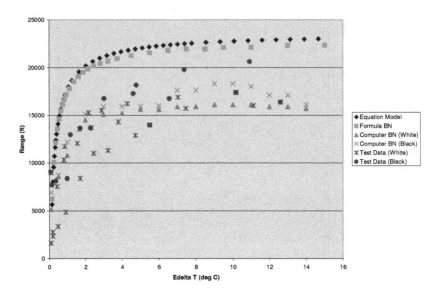

Figure F-13

FLIR Narrow FOV Model Comparison

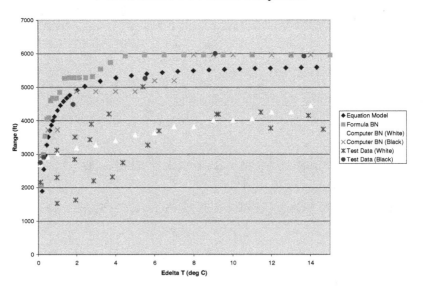

Figure F-14

FLIR WFOV Model Comparison

The average percent error for each model is presented in figure F-15.

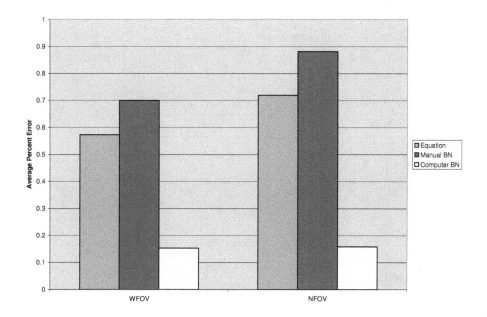

Figure F-15

FLIR Model Average Percent Error

Figure F-15 illustrates a major advantage the computer-generated model has over the other two models. The computer Bayesian network was able to locate the unmodeled variable "Polarity" resulting in two predictive curves for each field-of-view setting. These two curves represent the data much more accurately than the single curve of the other two models.

The time to complete the tasks required to build each model was recorded to the nearest minute. The results are presented in table F-1.

Table F-1

FLIR Model Construction Times

Action (minutes)	Equation-Based Model	Formula Bayesian Network	Computer Generated Bayesian Network
Develop Equations	70	70	-
Construct Model	35	38	4
Construct Neural Network	-	-	7
Computer Generation of Final Model	-	-	4
Total Time	105	108	15

As can be seen in table F-1, the computer generated model requires far less time to construct than either the equation-based model or formula Bayesian network.

Appendix G - Radar Cross Section Model

The radar cross section of an object is defined as that area which, when multiplied by the radar signal power density incident upon the target, yields a reflected power that, if radiated isotropically by the target, would result in a return back at the radar equal to that of the actual target [Masters, 1981b]. The radar cross section can be viewed as the effective area of a flat plate that would reflect the same amount of energy as the object measured in square meters [Knott et al, 1985]. For complex shapes, as radar energy strikes an object, the reflected energy is scattered in different directions depending on the shape of the object. For most cases, the radar transmitter, receiver and antenna are located at the same position so the amount of energy is measured as the amount directly reflected back to the antenna from the object. For simple geometric shapes such as cones, spheres, cylinders or plates, mathematical equations exist that can approximately calculate the radar cross section from the physical dimensions. Even for these simple shapes, the radar cross section depends strongly on the aspect angle between the target and the radar. It is also a function of the radar wavelength which is determined by the transmitter frequency. Both of these influences can be seen in the examples of figure G-1.

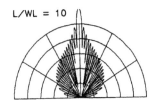 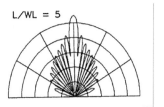

Figure G-1[9]

Theoretical Radar Cross Section for a Flat Plate

For these two examples, the ratio of the size of the plate (L) in relation to the wavelength of the radar is provided.

---

[9] From [Shaeffer, 1992]

Unfortunately, most real world targets are not simple geometric shapes. One method of predicting the radar cross section is to break down a complex shape into simple geometric shapes. This method may provide a rough prediction of the radar cross section across a sector of several degrees, but does a poor job of predicting the return at a specific aspect angle. This is due to the multi-bounce phenomena where radar energy reflected from one part of a complex object may be reflected one or more times off other parts of the object. These multiple reflections will then constructively or destructively interfere with each other on the return path to the radar receiver. An airplane is an example of a radar target with a complex shape. The radar cross section measurement at 10 GHz for a B-26 aircraft is shown in figure G-2.

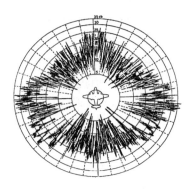

Figure G-2[10]

B-26 Radar Cross Section

As can be seen from figure G-2, the radar cross section measurement can change radically for small changes in aspect angle. Aspect angle can change because of motion of the target, the radar or both. The strong relationship between radar cross section and aspect angle explains why a target being tracked on radar may fade in and out at the edge of the maximum range of the radar as the aspect angle changes between radar transmission pulses. Thus, the performance of a

---

[10]From [Masters, 1981b]

radar detection system can fluctuate. For this reason, radar performance is usually described in terms of detection probability as a function of range.

Computer programs capable of predicting multi-bounce returns from complex objects have been developed by both universities and industry. These programs are complex and require special computer hardware to execute the extensive calculations. Many of these programs use optical methods to predict radar scattering. This is accomplished by breaking down complex shapes such as aircraft into a group of simple geometric shapes. An example of a geometric model of the A-10 aircraft using plates and cylinders is shown in figure G-3.

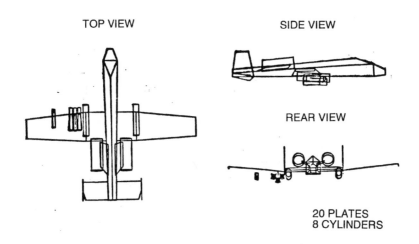

TOP VIEW SIDE VIEW

REAR VIEW

20 PLATES
8 CYLINDERS

Figure G-3[11]

A-10 Geometric Model

Although these models are capable of predicting the general areas of high and low signature returns, they can not exactly predict the radar cross section of the actual complex shape.

Developing an equation-based model is far beyond the scope of this research. The task of predicting radar cross section from complex shapes is also much too complex for human judgment. This effort is therefore limited to a machine constructed Bayesian network. A radar cross section model is created for the B-26 aircraft of figure G-2 using the Bayesian network

---

[11] From [Ryan, 1992]

construction software described in section 6.1. The model uses 40 bins and the normal probability option. The Bayesian network model is presented in figure G-4.

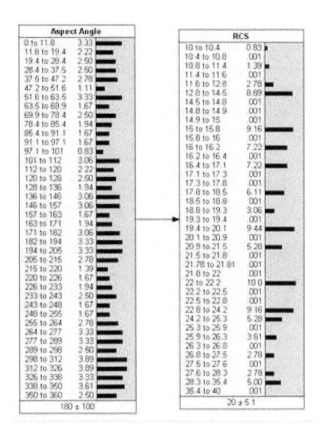

Figure G-4

B-26 Computer-Generated Radar Cross Section Model

The model predictions are compared to the test measurements in figure G-5.

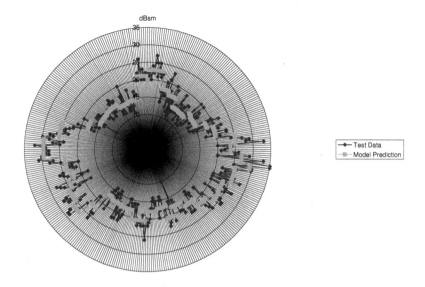

dBsm

Test Data
Model Prediction

Figure G-5

B-26 Radar Cross Section Comparison

The model prediction of figure G-5 shows the average cross section in square meters for the aircraft. The network also provides the range of values over which the radar cross section may vary as shown in figure G-6.

204

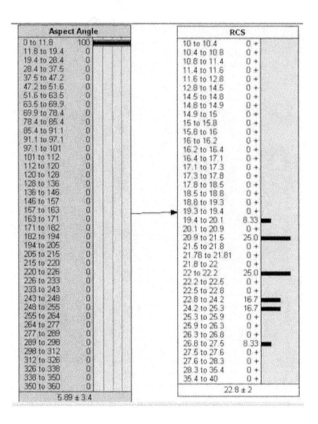

Figure G-6

RCS Bayesian Network Prediction

As can be seen from figure G-6, the radar cross section for 0 – 11.8 degrees is 22.8 ± 2 square meters.

Another advantage of Bayesian networks is that individual networks can be integrated with each other to solve complex problems. This is demonstrated by integrating the radar cross section model of figure G-6 with the computer-generated radar model constructed in appendix E to calculate the probability of detection of the radar against this target. The integrated model is shown in figure G-7.

205

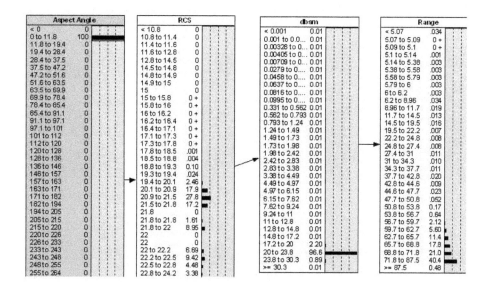

Figure G-7

Integrated Radar-RCS Bayesian Network Model

The probabilities in node "Range" are the individual probabilities for each bin at which the target will first be observed. These probabilities can be combined to find the cumulative probability of detection as shown in figure G-8.

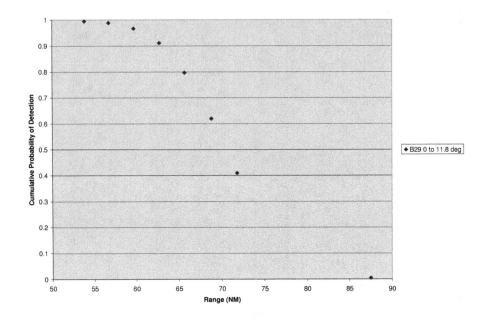

Figure G-8

Cumulative Probability of Detection for B-26 Aircraft at 0-11.8 Degrees Aspect Angle

Appendix H - Loan Application Process Simulation

The loan application process simulation is a virtual representation of the business processes used to approve or disapprove loan applications. This model is used to demonstrate business process reengineering simulation. The model looks at different screening processes used to route loan application forms through a loan application office. Times used for the processes are those provided with the original model. Due to privacy laws, actual loan application data was not available to test which method did the best job of screening the applications. Applications were generated by a random application generator for screening. Performance measurements are therefore limited to which processes can screen the most forms with the least amount of human labor.

The original simulation is shown in figure H-1.

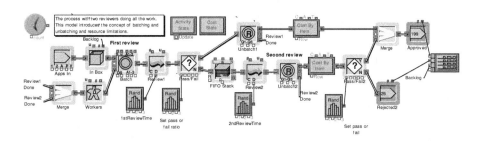

Figure H-1[12]

Loan Process Simulation

The process begins with applications coming into the office. There are two workers in the operation. Each application is given an initial review by a worker taking 12 to 16 minutes per review. If the application meets criteria set by the company for approval, the application is accepted. If it does not meet the initial screening requirements, the application is then forwarded for a second, more thorough review which takes 25 to 35 minutes after which a final

---

[12] Simulation provided with the Extend Version 5 software package. Author unknown.

determination of accept or reject is made. The simulation is run for a 40 hour work week with results presented in figure H-2.

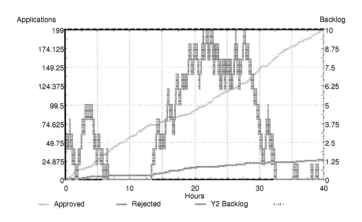

Figure H-2

Loan Screening Simulation Baseline Performance

From figure H-2, the baseline performance for the two workers is 225 applications processed with a maximum backlog of 10 applications. Worker utilization is 0.95 meaning that the two workers are processing applications 95% of the time and are idle awaiting applications 5% of the time.

The model of figure H-1 is next modified to investigate the improvements that could be achieved by automating part of the process. The modified model is presented in figure H-3. The file generator is modified to produce the same number of applications but adds six attributes to each: credit history, marital status, age, income, assets and debt. A random case generator was constructed to generate random cases that would generally follow demographic patterns and follow conventional knowledge concerning the likelihood of loan repayment. The first part of the file generator is shown in figure H-4.

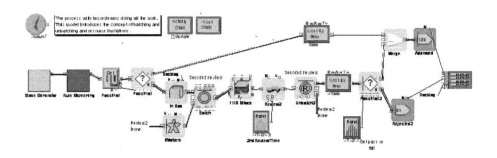

Figure H-3

Loan Application Rule Screening Modification

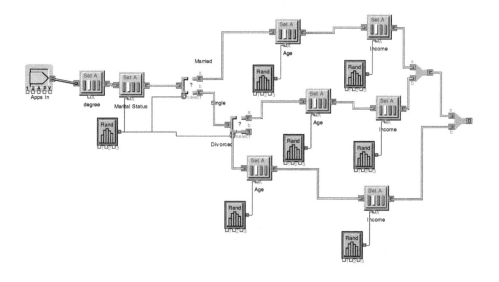

Figure H-4

Age, Marital Status and Income Probabilities

The applications are first assigned a marital status with 25% of applicants single, 45% married and 30% divorced. Single people are then routed and assigned a random age and income that on average is lower than that of married or divorced people. The process then continues in figure H-5.

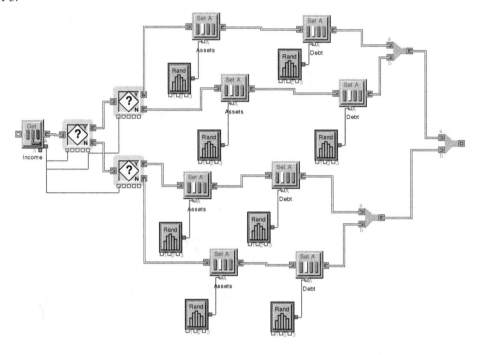

Figure H-5

Asset and Debt Probabilities

The model looks at income and then assigns a random value for assets and debts; however, the routing is set up such that those with higher incomes have more assets and fewer debts while those with lower incomes have generally lower assets and higher debts. The process then continues in figure H-6.

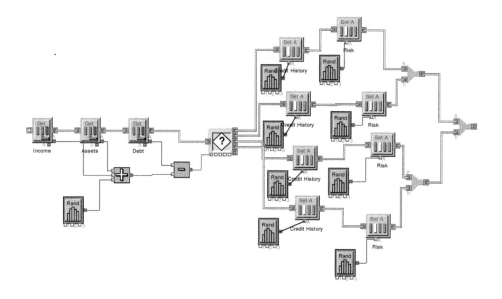

Figure H-6

Credit History and Risk Probabilities

The final process calculates the difference between the sum of income and assets and the debt. Applicants are then assigned a random credit history. Those with high positive values of the calculation are assigned a higher probability of a good credit history while those with a negative score are more likely to have a bad history. Intermediate scores are more likely to have intermediate credit ratings. The final assignment of risk is whether the loan was repaid. Those with good credit histories are more likely to repay than those with poor credit histories. Those with intermediate credit scores have values in the middle. The result is a wide combination of cases for review with trends that generally fit conventional wisdom with respect to the general population.

The first screening process done by workers in the baseline is replaced by a computer screening of the applications. The first part of the computer screening process is shown in figures H-7.

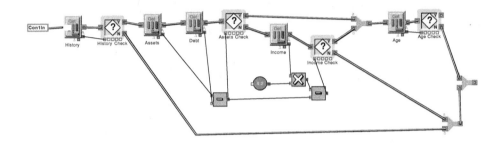

Figure H-7

First Loan Screening Process

The loan screening process is designed to flag any application that has any negative information. Applications which have no negative information are approved while those that contain one or more negative items are routed to the two workers for the second review. The process begins in figure H-4 by first checking credit histories and flagging any file with negative information. It then checks debt and assets to identify applications where debt is greater than assets. If debts are greater than assets, it flags any application where the difference is greater than 20% of current income. It then checks age and flags any application submitted by someone under 25. The process continues as shown in figure H-8.

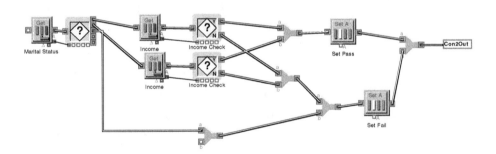

Figure H-8

Second Loan Screening Process

The process continues with a check to determine marital status. Those who are divorced have their applications flagged. Those who are single are flagged if they have less than $30,000 per year total income. Those who are married are flagged if they have less than $50,000 per year total income. This completes the screening process with all flagged files going to a second review with all non-flagged files approved without further action.

The simulation is then run to determine the impact on the loan application approval process. The results are presented in figure H-9.

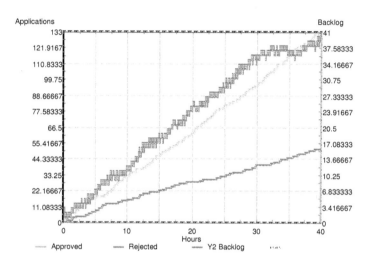

Figure H-9

Loan Application Rule Screening Results

The result is 184 applications processed with a maximum backlog of 41 applications and worker utilization rate of 100%. At first glance, the results are opposite to what one would expect. If the computer is doing the first screening of the applications and the workers now only have to do the second screening, why would it take longer? After further investigation, it was determined that the criteria used to screen the applications on the first screening by the computer was stringent resulting in a much higher number of files being sent to a second review than occurred during the original process. Because the second review is more thorough and takes longer, the end result was a higher workload for the two workers resulting in negative performance improvements by introducing automation into part of the process. Although there was no intent to have the results come out this way, it does show the importance of modeling and simulation in business process reengineering. It also demonstrates that adding automation does not always result in a better process.

A second computer screening process change is also evaluated to determine its impact on the loan application process. This approach is the same as described above except that instead of using the rule-based screening approach, the applications are now initially screened by a Bayesian network. The Bayesian network is presented in figure H-10.

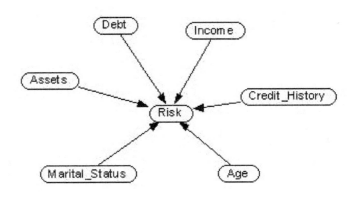

Figure H-10

Loan Application Screening Bayesian Network

The network structure was manually input and the probabilities were learned from a case file of previous loan applications. This file was generated by the same case file generator as is used to generate the loan applications for the simulation with the additional information of whether or not the loan was repaid. The simulation is the same as that shown in figure H-3 except that the computer screening of applications is changed to that shown in figure H-11.

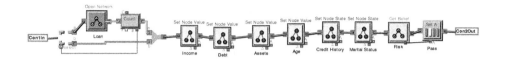

Figure H-11

Loan Screening Bayesian Network Integration

The network is opened and compiled on the first instance of an application. All six areas of interest are set into their respective node states. The program then retrieves the probability that someone with this set of factors will repay the loan. If this probability is 75% or higher, the loan is approved without further review. If below 75%, the application is sent to the workers for a second review.

This simulation is now run under the same set of conditions. The results are presented in figure H-12.

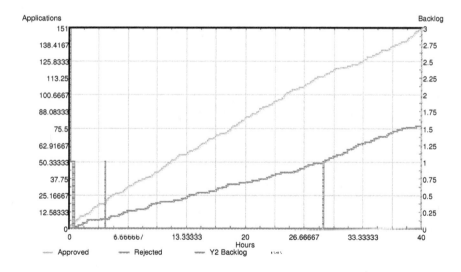

Figure H-12

Bayesian Network Screening Results

The result is 228 applications processed with a maximum backlog of 3 applications and a worker utilization rate of 0.58. This demonstrates a substantial improvement over the original process. Human labor has been reduced by nearly half. If the model is rerun with only one worker for the second review, the results of figure H-13 are obtained.

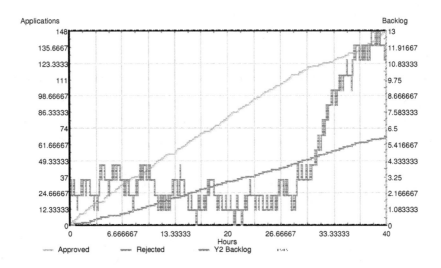

Figure H-13

Bayesian Network Screening with One Worker

The result is 225 loan applications screened with a maximum backlog of 13 applications and a worker utilization rate of 100%. This is the same number of applications screened with a maximum backlog increase of only 3 as compared to the original two review loan approval process.

The time to construct the screening portions of each model is presented in table H-1.

Table H-1

Computer Screening Process Construction Times

	Rule-Based Model (minutes)	Bayesian Network Model (minutes)
Construct Bayesian Network	-	15
Modify Loan Model	43	16
Total	43	31

Overall results demonstrate that the Bayesian network is far more effective in screening the application than the simplified rule-based screening process. Although the case files were not real loan applications, this simulation does show that Bayesian networks can learn from previous loan applications. The results should be no different on real applications as long as there the Bayesian network can adequately represent relevant patterns in the data.

Appendix I - Car Repair Process Simulation

The car repair process simulation is a virtual representation of the electrical shop at an automotive repair facility. The original simulation was constructed by the author and is shown in figure I-1.

Figure I-1

Car Electrical Repair Simulation

The process begins with cars coming into the service center. The "Car Generator" block creates a stream of cars that come into the service center with reported electrical problems. The model is set up to run on a standard 40 hour work week. Most cars are dropped off in the morning although a smaller number come in throughout each day. The "Car Generator" block is shown in figure I-2. The cars are assigned one of four electrical problems as follows: 33% are bad batteries, 22% have bad alternators, 25% have bad starters and 20% have short circuits in the electrical system. Once the fault has been determined, the results of three diagnostics are assigned as follows: does the battery hold a charge, are the headlights bright or dim with the engine running, and will the car jumpstart from another battery. In general, a car with a bad battery will not hold a charge, a car with a bad alternator will have dim headlights when the engine is running, a car with a bad starter will not jumpstart and a car with a short circuit can have any of the above symptoms.

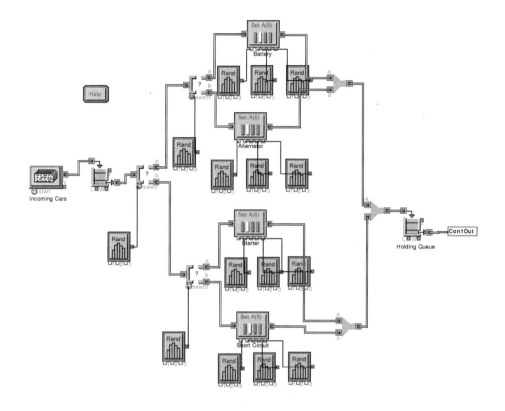

Figure I-2

Car Generator Block

The diagnostics are not foolproof in that some problems may create multiple symptoms and people dropping off their cars may report symptoms in error. Random noise of one to eight percent is added to each reported diagnostic to account for these variations. The cars are then routed to a holding area until a mechanic is available to repair them.

The electrical repair shop for this center has three service bays and three mechanics. The "Service Center" block is shown in figure I-3.

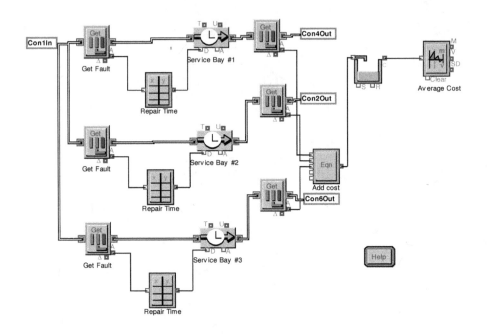

Figure I-3

Service Center Block

As the car enters the service area, the fault assigned to each car is retrieved and a time to repair is assigned. The car will be delayed in the service bay during the simulation for this amount of time. The three workers have different repair times for each repair. The first mechanic is the least experienced while the third mechanic is the most experienced. While it would take about the same amount of time for each mechanic to change a bad battery, it takes the more experienced mechanic much less time to isolate and repair a short circuit than the inexperienced one. The second mechanic and the other repairs have intermediate values assigned to each. Once repaired, cars are routed to the "Exit and Stats" block.

The "Exit and Stats" block records the results for each car and calculates the statistics over a 52 week operating period. The "Exit and Stats" block is shown in figure I-4.

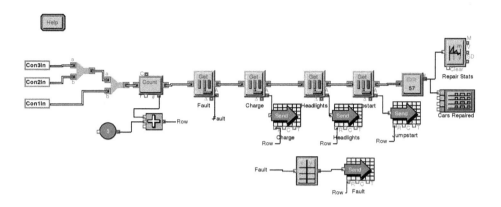

Figure I-4

Exit and Statistics Block

The fault along with each of the three diagnostics is written to a Microsoft Excel spreadsheet. The number of cars repaired per week and the amount of money collected per week are also calculated. The repair process is completed when cars exit the repair shop and are returned to their owners.

A second model was developed by modifying the initial car repair model. Whereas the baseline simulation routes cars to the mechanics on a first in, first out basis this second simulation uses a rule-based approach to sort the cars into four waiting areas. Cars are sorted by suspected fault based on diagnostic indications. The sorting modification, which is added to the "Service Center" block, is shown in figure I-5.

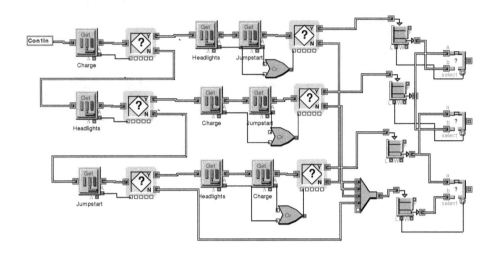

Figure I-5

Rule-Based Sorting

The rules are as follows: fails to hold charge only goes to area 1, dim headlights only goes to area 2, fails to jumpstart only goes to area 3 and everything else goes to area 4. This is designed to put cars with bad batteries in area 1, bad alternators in area 2, bad starters in area 3 and short circuits in area 4. This ordering is the same as the degree of difficulty of each repair. Mechanic 1, the least experienced mechanic, draws cars from areas 1 and 2; mechanic 2 draws from areas 2 and 3; and mechanic 3 draws from areas 3 and 4. This sorting is designed to send the most difficult repairs to the most experienced mechanic, who can repair them the most quickly. Drawing from two different holding areas is designed to ensure that no one is idle awaiting a car to repair.

A third variation of the model was created using an influence diagram that makes a decision as to which holding area the cars are routed based on maximizing the profit to the repair shop. The influence diagram relations are a naïve Bayes model and were manually created. Probabilities were learned from the data captured in the Excel spreadsheet as an output of the baseline model. The influence diagram is presented in figure I-6.

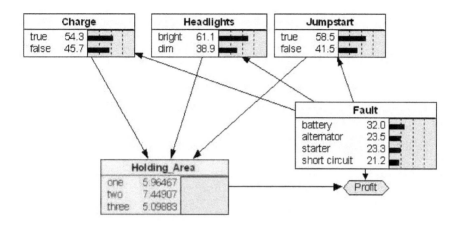

Figure I-6

Influence Diagram Decision Model

The network communicates with the simulation via the blocks shown in figure I-7.

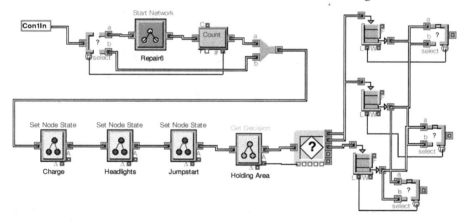

Figure I-7

Car Repair Network Integration

The network is opened and compiled once the first time a vehicle comes through. The indicators are applied as evidence to their respective nodes in the influence diagram. Based on the diagnosis of the fault, the influence diagram then makes a decision as to which holding area the car is routed based on optimizing the profit.

The three different simulations were run for 52 weeks each. Repairs are billed on a fixed-price basis dependent on the actual fault. The results are presented in figure I-8.

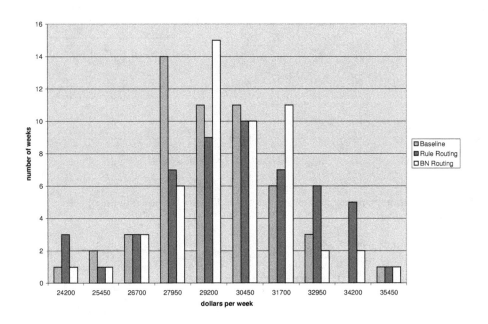

Figure I-8

Car Repair Simulation Comparison

As can be seen in figure I-8, all methods result in Gaussian distributions. Routing the cars based on an initial diagnosis results in improved revenue for the repair shop over the first in, first out baseline. Rule-based routing on average was 19.3% higher than the baseline data while the influence diagram routing was 24.0% higher. Influence diagram routing was more effective than rule-based routing because it had a higher rate of correctly diagnosing the problem. The rule based approach used a simple screening to attempt to match the symptom most associated with a fault to that fault while diagnosing anything else as a short circuit. The influence diagram learned from previous data which symptoms were associated with which faults. It was therefore more successful in diagnosing faults from a noisy data set.

The time required to construct the diagnostic models is presented in table I-1.

Table I-1

Car Diagnostic Model Construction Times

Actions (minutes)	Rule-Based Model (minutes)	Influence Diagram Model (minutes)
Build Influence Diagram	-	8
Modify Baseline Model	34	22
Total Time	34	30

Appendix J - Home Heating System Simulation

The home heating system simulation is a virtual representation of a house with an oil furnace in the Washington DC area during the month of January. The model represents a single story, 1875 square foot house with three doors and ten windows. The model contains handbook insulation values for the materials selected. The baseline model is shown in figure J-1.

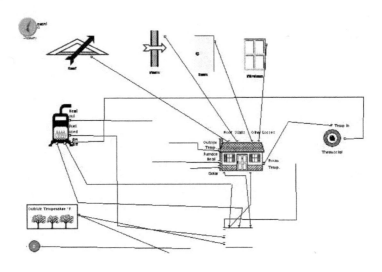

Figure J-1[13]

Home Heating System Model

The model consists of a house which is subjected to heat losses and gains. Losses occur through the roof, walls, doors and windows shown at the top. The amount of heat loss is a function of the outside air temperature which varies over each 24 hour period. The air temperature is a sinusoidal function that varies between the high and low temperature of each day. Heat gains are provided by the furnace element on the left side of the figure which is controlled by a thermostat at the right side. The baseline thermostat model is shown in figure J-2.

---

[13] Model modified by the author to discrete operation from a model provided with the Extend M&S package. Original author is unknown.

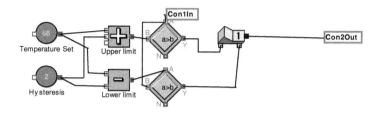

Figure J-2

Baseline Thermostat Model

The thermostat is set at 68 degrees F and has a 2 degree hysteresis (furnace turns on at 66 degrees and off at 70 degrees).

    The model is executed over a 24 hour period simulating the interactions of the house and furnace with the outside environment. The temperature data are provided in figure J-3.

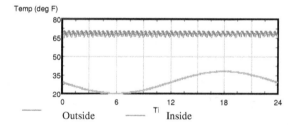

Figure J-3

Baseline Temperature Data

The blue lower line displays the outside air temperature which varies from a low of about 20 degrees F to a high of about 40 degrees F. The upper line displays the inside air temperature which varies between 66 and 70 degrees F as the furnace cycles on and off throughout the day. The fuel consumption for one 24 hour period is shown in figure J-4.

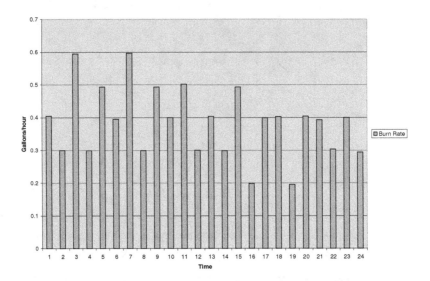

Figure J-4

Baseline Fuel Consumption Rate

The consumption varies over the period with outside air temperature. Consumption is higher in the first half of the period corresponding to the low temperature portion of the outside air temperature curve shown in figure J-3 and lower in the second half where outside temperatures are higher. Fuel consumption for 24 hours under these conditions is 9.4 gallons.

The model is then modified with a programmable thermostat. The thermostat can change the temperature setting four times per day. The modified thermostat model is shown in figure J-5.

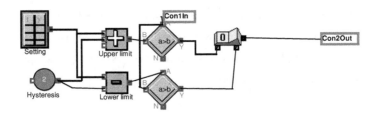

Figure J-5

Programmable Thermostat

The thermostat is set as follows: 62 degrees from 10 PM to 6 AM, 69 degrees from 6 AM to 9 AM, 66 degrees from 9 AM to 5 PM and 68 degrees from 5 PM to 10 PM. The model is then run over a 24 hour period to simulate the interactions of the house with the outside environment with the results shown in figure J-6.

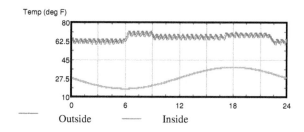

Figure J-6

Temperatures with Programmable Thermostat

Figure J-6 shows the four changes to the inside air temperature as the result of inserting the programmable thermostat.

The baseline model is modified a second time using a Bayesian network model of a programmable thermostat. The temperature settings are the same as those used for the programmable thermostat described above. The Bayesian network model is shown in figure J-7.

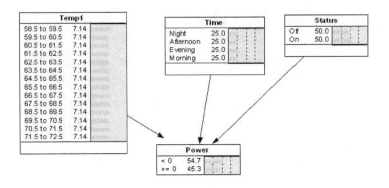

Temp1	
58.5 to 59.5	7.14
59.5 to 60.5	7.14
60.5 to 61.5	7.14
61.5 to 62.5	7.14
62.5 to 63.5	7.14
63.5 to 64.5	7.14
64.5 to 65.5	7.14
65.5 to 66.5	7.14
66.5 to 67.5	7.14
67.5 to 68.5	7.14
68.5 to 69.5	7.14
69.5 to 70.5	7.14
70.5 to 71.5	7.14
71.5 to 72.5	7.14

Time	
Night	25.0
Afternoon	25.0
Evening	25.0
Morning	25.0

Status	
Off	50.0
On	50.0

Power	
< 0	54.7
>= 0	45.3

Figure J-7

Bayesian Network Programmable Thermostat

The network has three root nodes with "Temp1" for the current inside air temperature, "Time" for the time of day with each state set to a value of the thermostat setting for that period and "Status" for the current status of the furnace (on or off). The "Power" node uses a formula to determine if the furnace should be switched on or off with "<0" being the switch on command and ">=0" being the switch off command. The thermostat of figure 6 is then integrated into the baseline model by replacing the thermostat of figure J-2 with the one shown in figure J-7. The integration blocks are shown in figure J-8. In the first row of blocks, the model first determines which of the four time periods it is in based on the time of the simulation. The model then sets the current inside air temperature and the status of the furnace. In the second row, the Bayesian network is opened and compiled on the first iteration of the simulation. The three states of nodes "Temp1", "Time" and "Status" are then set in the Bayesian network. The most likely state of node "Power" is retrieved which is then used to switch the furnace on or off in the simulation. The simulation is then run over a 24 hour period with results shown in figure J-9.

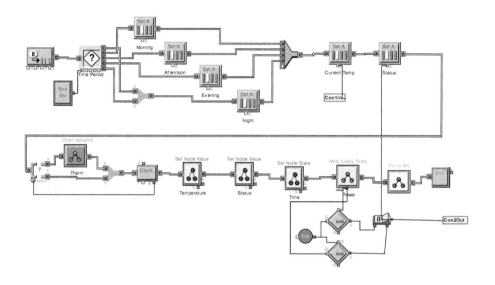

Figure J-8

Bayesian Network Thermostat Integration

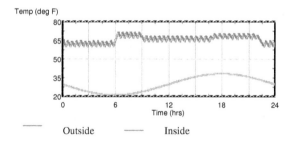

Figure J-9

Bayesian Network Thermostat Simulation

As expected, the result is a plot very similar to the results of the equation programmable thermostat simulation.

Each model is then set to use a random normal distribution for the high and low temperature of each 24 hour period using the mean and standard deviation for temperature in the

232

Washington DC area for the month of January. The simulations are then run 30 times to calculate the fuel consumption distribution. The results are shown in figure J-10.

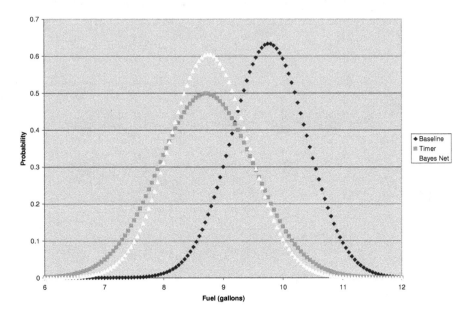

Figure J-10

Daily Fuel Consumption Comparison

The model timer and the Bayesian network timer simulations produced nearly the same result. Both simulations that used timer thermostat models showed an average daily fuel consumption of approximately one gallon less than the fixed thermostat of the baseline simulation.

The time required to create the two timed thermostat models is shown in table J-1.

Table J-1

Thermostat Model Construction Times

Actions (minutes)	Equation Timer Model	Bayesian Network Timer Model
Model Construction	28	17
Model Integration	5	24
Total	33	41

Both timer thermostats used equations in the model construction. As expected, both simulations provided about the same results and took similar times to build.

Appendix K - Commuter Simulation

    The commuter simulation is a virtual representation of driving an automobile from Defense Acquisition University at Fort Belvoir, VA to the author's home in Oakton, VA during rush hour traffic. This is a difficult model to build using conventional equation-based techniques because of a number of uncontrollable random factors such as traffic density, accidents and weather which influence the outcome. It also demonstrates the capability of the derivative method to discretize large numbers of continuous variables simultaneously. The baseline equation model is shown in figure K-1.

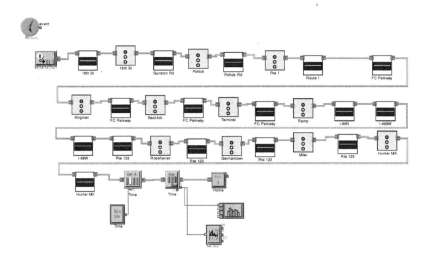

Figure K-1

Equation Commuter Model

The model consists of all of the route segments and traffic lights along the entire trip. On route segments, the model assumes the car will travel at the average speed limit for that leg. The traffic lights are empirical probability tables that generate a random delay based on measured values of the red and green cycle times of each light. A yellow light is considered part of the green cycle time since it is permissible to go through a yellow light. As an example, if a light has a 25 second time of green and then a 25 second time of red, the probability would be 0.5 for zero delay and 0.1 each for 5, 10, 15, 20 and 25 second delays. Although this ignores timing of multiple lights to coordinate traffic flow, only three lights were sequenced in the direction of travel for this model. Rush hour traffic is usually backed up on those segments negating any impact of the sequencing. The car moves from top to bottom through all legs and lights in figure K-1 and the total time is calculated. A Monte Carlo simulation is run 10,000 times to generate a probability distribution of the total driving time. The output distribution is shown in figure K-2.

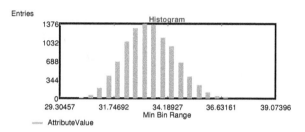

Figure K-2

Monte Carlo Distribution for Commute Simulation

This results in an approximately Gaussian distribution with mean of 33.60 and standard deviation of 1.30 minutes.

A manually constructed Bayesian network model was created using the same data and assumptions as the equation-based model of figure K-1. The Bayesian network model is presented in figure K-3.

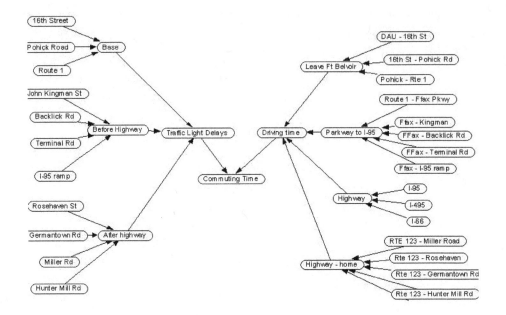

Figure K-3

Manual Bayesian Network Commute Model

The model of figure K-3 is constructed with all traffic light delays on the left side and all leg driving times on the right. The center nodes use an equation to sum the inputs of the driving and traffic light delay times. The "Commuting Time" node in the center is the sum of all the nodes along either side. The Bayesian network also calculates a Gaussian distribution shown in figure K-4.

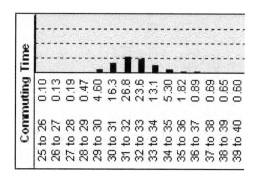

Figure K-4

Manual Bayesian Network Commute Time Distribution

This distribution has a mean of 32.9 and standard deviation of 4.2 minutes. The mean is very similar to that obtained with the Monte Carlo simulation shown in figure K-2. However, the standard deviation is wider and more accurately reflects the total possible spread of values. This demonstrates an advantage of using Bayesian networks in complex systems with multiple random variables. The probability of all traffic lights being green is 0.0027 while the probability of arriving at all signals just as they turn red is 5.69E-13. While the Bayesian network was able to calculate the total probability range in one computer cycle, a Monte Carlo simulation of the equation-based model would have to run 1.76E12 samples in order to theoretically capture the rare event of arriving at all the traffic lights as they turn red.

A third model was constructed using measured travel times over the route described above. Additional information including the day of the week the trip was made and the start time of the trip were also recorded. BN PowerConstructor was used to determine structural relationships which are shown in figure K-5. A relationship was found between the day and time of departure and other nodes of the network. BN Builder was then used to reconstruct the model using the software described in section 6.1. This required the simultaneous discretization of 29 continuous nodes. The distribution of node "Total" is shown in figure K-6.

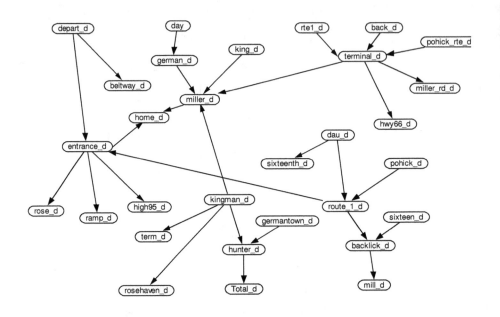

Figure K-5

Computer-Constructed Commute Model

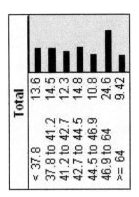

Figure K-6

Computer Bayesian Network Commute Time Distribution

This simulation had a non-Gaussian distribution with a mean travel time of 47.6 minutes.

The error percentage from the mean value of each prediction to the measured mean of 46.46 minutes is presented in figure K-7.

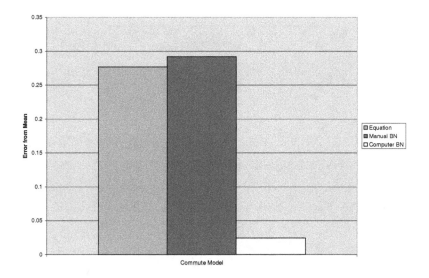

Figure K-7

Commute Simulation Error Comparison

This shows that many factors such as traffic, weather, accidents, etc. that are difficult to quantify when constructing a model prior to collection of data do have a significant impact on the outcome.

The equation model of figure K-1 was updated to include the measured data used to create the computer generated model. The traffic light delays remain the same but the leg times are replaced with a normal distribution of the measured data for each route leg. A Monte Carlo simulation is then run with 10,000 samples with the results shown in figure K-8.

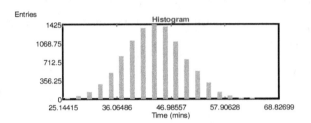

Figure K-8

Updated Equation Monte Carlo Simulation

This results in a Gaussian distribution with a mean commute time of 44.83 and standard deviation of 5.98 minutes. The Bayesian network model of figure K-3 was also updated with the data collected from driving the route. This data was used to update all root nodes (both leg times and light delays). The updated network has a Gaussian distribution of driving times presented in figure K-9.

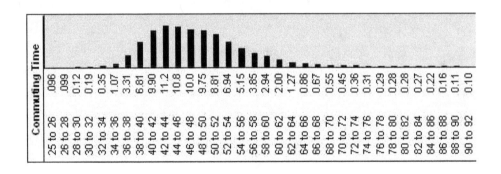

Figure K-9

Updated Bayesian Network Commute Time Distribution

This distribution is Gaussian with a mean driving time of 48.7 and a standard deviation of 9.2 minutes.

The updated models were tested using three additional commute times which were not used in the probability updates. The percent difference of the prediction means from the measured values are compared in figure K-10.

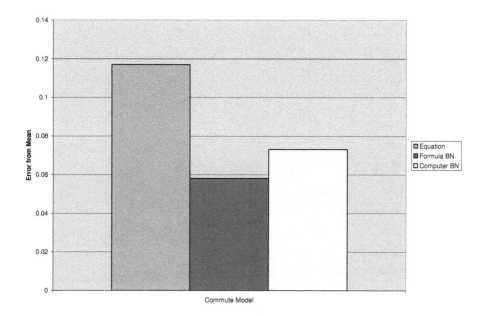

Figure K-10

Commute Model Error Comparison

Even when updated with the measurement data, both of the Bayesian networks had lower errors than the equation-based model. The times required to construct each of the three models in presented in table K-1.

This particular example demonstrates two additional advantages of using Bayesian networks. First, the manually created network is constructed such that all times are independent of each other (each input node has a series of arcs that goes from that node to the total time node without going through another input node). This network can be used to refine the prediction of travel time as the route is traversed. As the actual travel time for each leg or delay time for each light is entered, the mean travel time is updated and the spread of possible values is reduced.

This is demonstrated in figure K-11 where approximately two-thirds of the route is traversed with the times entered.

Table K-1

Commuter Model Construction Times

Action (minutes)	Equation Model	Manual Bayesian Network	Machine Learning Bayesian Network
Lay Out Route and Calculate Distances	40	40	-
Measure Light Times and Calculate Probabilities	95	95	-
Construct Model	74	107	6
Computer Generation of Final Model	-	-	3
Total Time	209	242	9

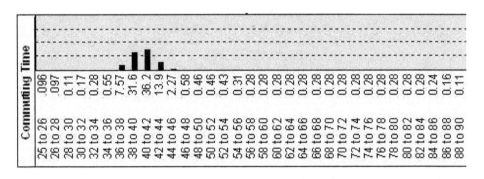

Figure K-11

Commute Time Prediction at Halfway Point

Comparing figure K-11 with figure K-9, one can see that the new mean travel time prediction is now 42.2 and that the distribution spread has been significantly reduced. Although it is possible to do the same prediction with the equation-based model, the route and light values would have to be changed from probability distributions to constants and the Monte Carlo simulation run for another 10,000 iterations. This is certainly not feasible for real time updates of the travel time predictions.

A second advantage is that Bayesian networks can find unmodeled relationships which influence outcomes. The computer generated network in figure K-5 found that both the day of the week and the time of departure influenced the network time variables. These two variables are the only controllable variables that are inputs to the models. If one is only interested in the probability distribution of total driving time, the network structure of figure K-12 is obtained from BN PowerConstructor if data values "day" and "time" are declared root nodes and "Total" is declared a leaf node.

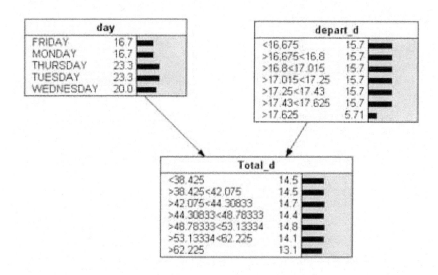

Figure K-12

Simplified Commute Network Structure

Using BN PowerConstructor ensures that there are in fact relations between "day" and "Total" and "time" and "Total". There are only 25 test measurements so that every cell does not have test data for a probability calculation. The neural network option can be used to predict times for day/time combinations that were not measured. The neural network is shown in figure K-13.

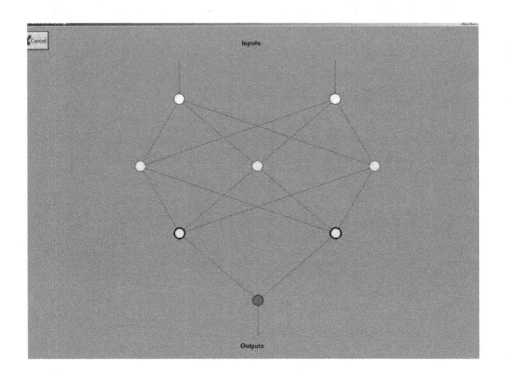

Figure K-13

Commute Neural Network

The three element data set, the structure of figure K-12 and neural network of figure K-13 are used as inputs to the BN Builder software with the output network shown in figure K-14.

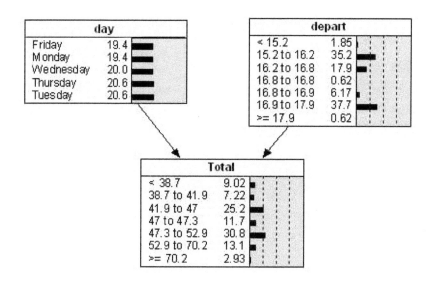

Figure K-14

Day/Time Commute Model

This model can now make a probabilistic prediction of commute time based on inputs "day" and "time". For example, if it is Tuesday the model predicts that the average commute time is 52 minutes and one should leave between 4:48 and 4:54 PM to minimize driving time. If one wishes to be 90% confident of arriving home by 6 PM on Wednesday, then departure can be no later than 5:05 PM. The same types of predictions are not possible using an equation-based model without dramatically increasing the complexity of the model. Although the light times would remain the same, each leg of the journey and departure time period would need a separate random probability distribution for each day of the week. This particular example shows strong advantages of using Bayesian networks when modeling complex, ill-defined systems where the outputs will vary over a range of values.

L. Integrated Radar/RCS and Air Defense Models

L.1 Integrated Radar Tracking Model

Appendix G provides an example of a radar cross section model of the B-26 aircraft shown in figure G-6. Appendix E provides an equation-based model of the F-16 radar shown in figures E-1 through E-3. The equation-based radar model was adjusted with additional losses so that the model results validate with the test data. In this example, it is assumed that the validated radar model now already exists. The aircraft will be tracked by the radar requiring a motion model. Because object motion is a well defined problem for which validated equations exist, an equation-based model is used for motion. The motion model is shown in figure L-1.

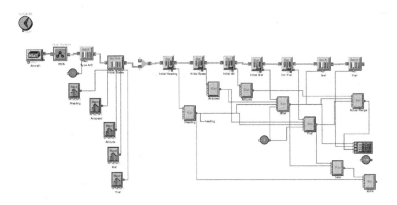

Figure L-1

Integrated Radar/Target Motion Model

The first blocks on the left open the Bayesian network and set heading, airspeed, altitude and initial position of the target. The rest of the blocks set those conditions as attributes in the simulation and calculates the movement of the aircraft between radar scans and the aspect angle of the target.

The RCS of the target, as previously described, is a complex problem for which good equation-based methods do not exist. A computer constructed Bayesian network is chosen for this element of the model. The Bayesian network model integration is shown in figure L-2.

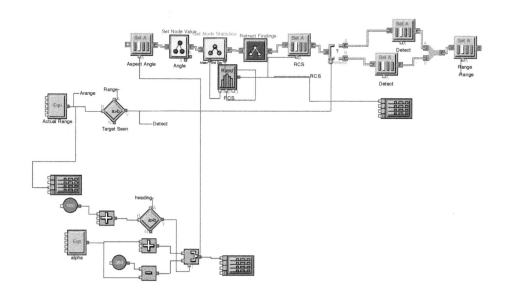

Figure L-2

Integrated Radar/Target Bayesian Network Integration

The upper blocks set the aspect angle into the Bayesian network and retrieve the mean and standard deviation of the radar cross section. The model then generates an RCS value for this sweep by selecting a random number from a normal distribution defined by the returned mean and standard deviation. This number is sent to the model of figure E-1 of appendix E in place of the RCS input block. The model then calculates the maximum range of the radar for this RCS value. The center blocks compare the maximum range of the radar with the current range of the target. If the actual range is less, the target is detected. Target detection is then set in the far right blocks of the model. The lowest blocks of the model correct the aspect angle to a value of 0 to 360 degrees if it is negative. The parameters of the simulation are captured in the blocks shown in figure L-3.

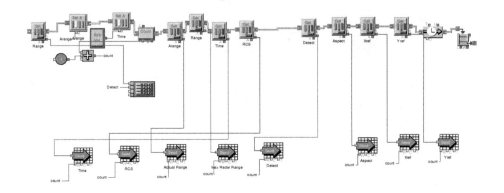

Figure L-3

Integrated Radar/Target Simulation Recording

The blocks of figure L-3 capture all the data parameters in an Excel spreadsheet. The last two blocks at the far right delay the simulation for 2.5 seconds, the time between radar sweeps. The aircraft is then routed back to the motion model for calculations of position and aspect angle for the next sweep. This iterative looping continues until the end of the simulation.

The simulation is now run over a period 16.67 minutes such that the aircraft moves toward the radar. The radar does not move and is positioned at coordinates X=0, Y=0. The F-16 radar track of the B-29 is shown in figure L-4. The simulation provides a realistic simulation of radar operation. The target initially returns a few detections, then transitions to about half detections and half non detections and then finally moves in to a closer range where all returns are detections. This real world phenomenon where a target fades in and out near the outer detection range is known as target scintillation.

A second target model was created for a 1/15 scale model of the Boeing 737 commercial aircraft. The radar cross section measurement at 10 GHz is shown in figure L-5.

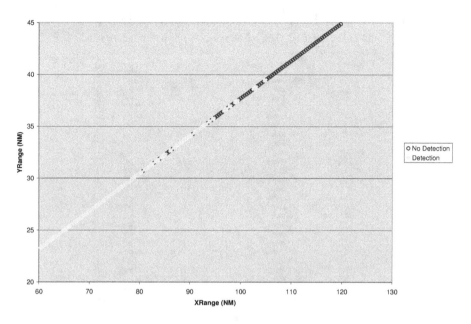

Figure L-4

Integrated Radar/B-26 Tracking Simulation

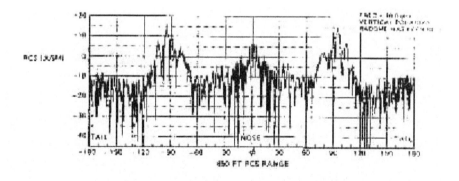

Figure L-5[14]

RCS Measurement of a 1/15 Scale B-737 Aircraft

---

[14] Knott, Eugene, Shaeffer, John and Tuley, Michael, Radar Cross Section, Artech House Inc., Norwood MA, 1985, p. 181.

The data of figure L-5 was used to create a Bayesian Network model of the aircraft RCS using the software described in section 6.1. The RCS model is shown in figure L-6.

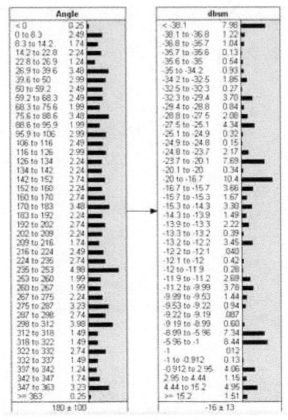

Figure L-6

1/15 Scale B-737 Aircraft RCS Model

A comparison of the model predictions with the test data is shown in figure L-7.

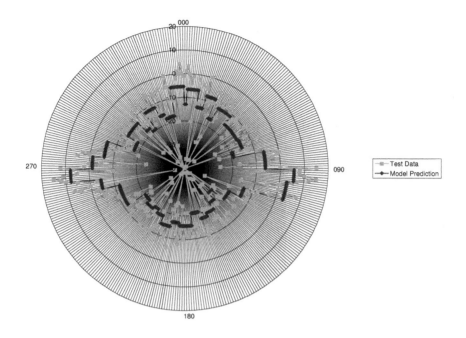

Figure L-7

1/15 Scale B-737 Aircraft RCS Model/Test Comparison

This model was then run in the integrated radar/RCS model over a different track to determine the detection range. The results are shown in figure L-8.

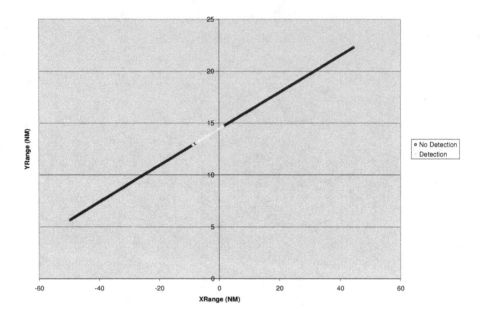

Figure L-8

Integrated Radar/1/15 Scale B-737 Tracking Simulation

As can be seen in figure L-8, the detection range of the much smaller model is much less than the B-26 aircraft shown in figure L-4. The RCS value determined by the Bayesian network model is shown as a function of the aspect angle is shown in figure L-9.

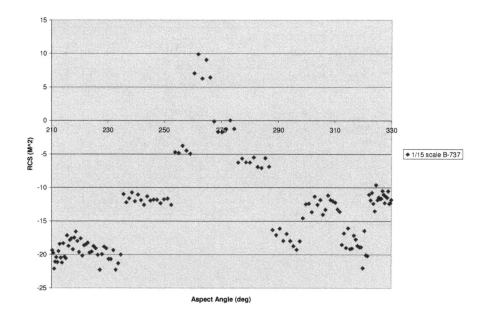

Figure L-9

1/15 Scale B-737 RCS with Aspect Angle

As can be seen in figures L-7, L-8 and L-9, the primary reason the radar can see the target as it passes the radar is not a function of target range, but the higher RCS value correctly modeled by the Bayesian network as the aspect angle of the target changes with target motion. As the aircraft passes so that the radar is looking at the left side of the aircraft, the RCS increases dramatically just aft of the wing as shown in figures L-7 and L-9 resulting in target detection. As the aircraft passes on so that the aspect angle moves into the left rear quarter, the RCS falls and the detection is lost even though the aircraft continues to move closer to the radar.

L.2 Integrated Air Defense Decision Model

A final integrated model was created to demonstrate how influence diagrams can be tested and improved by allowing them to interact with a simulation. The combat identification Bayesian network of figure L-10 was used as the starting point.

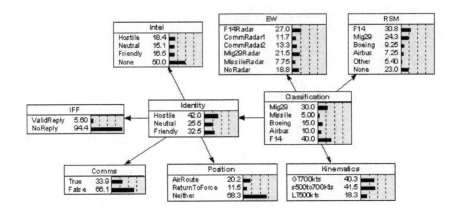

Figure L-10

Combat Identification Network[15]

This Bayesian network uses seven attributes of a target to identify the type of air vehicle and determine whether it is hostile, neutral or friendly. These attributes include the speed (Kinematics), Position relative to a commercial airway or specified return to force procedures, radio communications with the target (Comms), whether the contact has an Identify Friend or Foe (IFF) transponder turned on, Intelligence about the target, electronic signals (EW) being emitted by radar onboard the target, and any change in the frequency of the radar return caused by a Doppler shift of the radar energy bouncing off the front of the engine (RSM). This network was expanded into an influence diagram by adding a utility node and a decision node for recommending firing or not firing a weapon at a target that is detected by the radar. The radar cross sections for three types of aircraft were then added for use with the equation-based radar model as shown in figure L-11.

---

[15] From [Laskey and Laskey, 2002]. Used with permission.

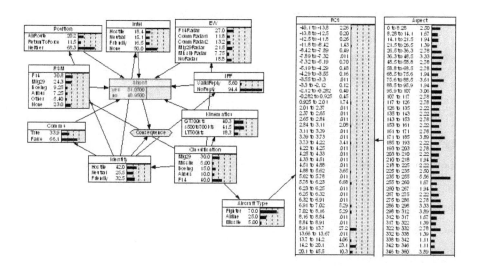

Figure L-11

Baseline Air Defense Influence Diagram

The decision whether to shot at an incoming target is a function of the probability that a target is hostile and the utilities assigned in figure L-12.

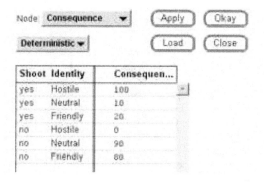

Figure L-12

Shoot Decision Utility Values

256

As can be seen from figure, the most favorable outcomes are to shoot at a hostile target while not shooting at neutral and friendly targets. The most unfavorable outcomes are to not shoot at a hostile target while shooting at neutral and friendly targets.

The influence diagram of figure L-11 was then integrated into the equation-based model of figure L-1 through L-3. The equation-based model first determined if the target would be detected by the radar. If the target was detected, the values for nodes "Kinematics", "Position" and "RSM" were determined by the speed, position and aspect angle of the target as calculated by the equation-based simulation. Nodes "EW", "Comms" and "IFF" were randomly determined by a random number generator set to realistic values for each type of aircraft. Random errors were added to this data to test how well the network would work under realistic conditions. For example, on the MIG-29 EW node, there was a 50% chance the radar was on and identified as MIG-29, a 30% chance it was either not on or not picked up, and a 5% chance that it was misclassified as each one of the other four options.

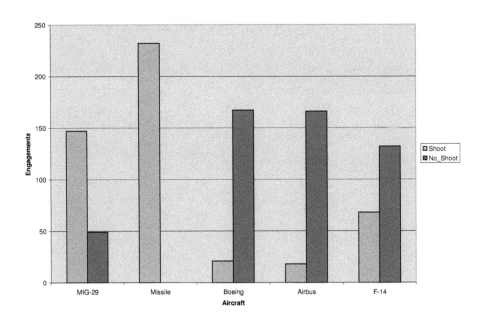

Figure L-13

Baseline Air Defense Influence Diagram Results

The model was then run for 1000 aircraft engagements to test how well the network could correctly identify the type aircraft and make a correct decision whether to recommend firing at the contact. The results of these tests are shown in figure L-13. The results of the simulation showed that the network had problems both correctly identifying the type of target and in the underlying policy of determining whether a target is hostile. Although the network successfully identified and recommended firing at all incoming missiles, it also recommended firing at approximately 10% of the neutral targets and a third of friendly aircraft. This was caused by the conditional probabilities identifying the type of target. The network also recommended firing at 75% of the MIG-29 aircraft. This demonstrates a policy error in the logic of the original network. Although one can define any inbound missile as a hostile threat, one would not fire at an aircraft just because it is a specific type.

The decision structure of figure L-11 was first modified by adding arcs from node "Identity" to nodes "EW" and "Kinematics". A new policy that a MIG-29 flying at the radar at speeds greater than 500 knots and with their radar on is considered to be hostile. Otherwise, these aircraft are neutral. The influence diagram was cleared of all probabilities and utility values. The equation-based model generated random aircraft engagements, but was modified to randomly make the shoot decision and then determine the utility of that decision. Once all node states for the influence diagram were set, the probabilities and utility values were updated. This allowed the influence diagram to learn by evaluating the outcome of each simulated engagement. In this way, the influence diagram determined the unconditional and conditional probabilities for each natural node and which decisions had the most favorable outcomes. The simulation was run 10,000 times to determine the probabilities, utilities and optimal decisions.

The updated network was then integrated with the original simulation and run for another 1000 engagements. The results are provided in figure L-14,

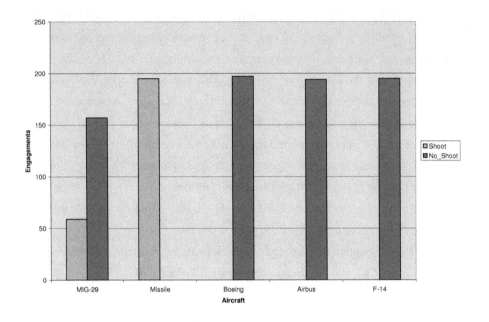

Figure L-14

Simulation-Trained Air Defense Influence Diagram Results

The network with revised decision policy and probabilities learned from simulation provided much better results. The network correctly identified all incoming missiles and recommended a decision to shoot. The network was now able to recognized 100% of all types of neutral and friendly aircraft and recommended not shooting in all cases. The MIG-29 was also correctly identified and a shoot decision was recommended in the 27% of cases were the aircraft made a high speed run at the radar position with its radar on.

This example demonstrates the advantages of integrating equation-based models with influence diagram in simulations. The equation-based portion of the model can be used to determine the input states of the nodes of the influence diagram. In this example, the model generated random starting points, headings and airspeeds for each target. The equations of motion and the capabilities of the radar determined at what position and aspect angle the target appeared on the radar. This information then determined whether the target was detected. This approach provides more realistic input scenarios than using random number inputs for the variables. The influence diagram can then make decisions which are fed back to and effect the simulation. In this case, the returned decision was whether to shoot at the radar contact. Although not implemented in this example, an equation or Bayesian network model of a weapon could be implemented to continue the simulation. Integrated simulation can also be used to test decision policy and to determine the probability tables of natural nodes. For complex decisions, integrated simulation can show whether the desired results are obtained for the given inputs. This is particularly important in cases were inputs are either incorrect or not available. Provided the simulation is representative of the real world, an influence diagram can be trained and tested prior to implementation.

Appendix M - Robotic Vehicle Models

The robotic vehicle models are a virtual representation of a robotic car constructed from the Lego™ MindStorms® Robotics Invention System. The baseline vehicle is shown in figure M-1.

Figure M-1
Baseline Robotic Vehicle

An equation-based model was created to predict the speed of the vehicle. The selection and input sections of the model are presented in figure M-2.

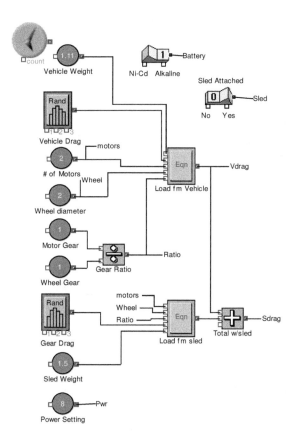

Figure M-2

Robotic Vehicle Equation Model Inputs

The model elements on the left side are the input variables. The top two switches allow the model to run using either alkaline or nickel-cadmium batteries and to attach or detach a 1.5 pound sled for towing speed prediction. The calculations in figure M-2 find the vehicle drag and the sled drag.

The motor portion of the model is presented in figure M-3.

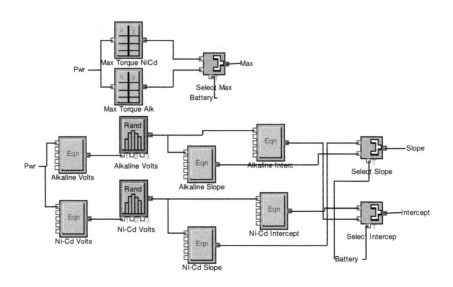

Figure M-3

Robotic Vehicle Equation Motor Model

The top portion of figure M-3 calculates the maximum torque output of the motor based on the type of battery selected. The lower portion of the model calculates parameters used in the RPM calculations. The final portion of the model is shown in figure M-4.

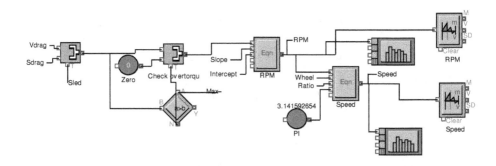

Figure M-4

Robotic Vehicle Equation Speed Model

The first portion of figure M-4 calculates the RPM of the motor. If the torque load is greater than the maximum torque from the motors, the RPM is zero. The final portion calculates the final vehicle speed based on the RPM, gear ratio and wheel size.

Equation model predictions were run for three different configurations. The kit used to construct the car has some variation among the parts. The test area carpet also did not have uniform friction over the surface. It was assessed that these values could vary up to 10% in value. These elements were modeled as normally distributed random variables with the mean set to the average value and standard deviation of 3.3% of the mean. Battery voltage accounts for the largest variation in system performance. Battery voltage was tested with the results presented in figure M-5.

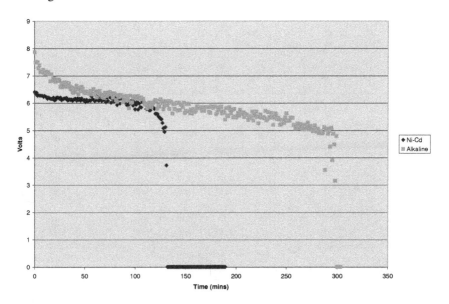

Figure M-5

Battery Test Data

The battery voltage was modeled as a random variable based on a normal distribution of the test data.

System performance predictions were made using Monte Carlo simulation varying the gear drag, sled drag and battery voltage over the range of values. The prediction probability distribution was then compared to a normal distribution of test data taken for four different configurations. These comparisons are presented in figures M-6 through M-9.

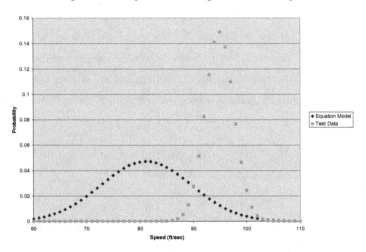

Figure M-6

Ni-Cd Battery, Power 8, No Sled Performance

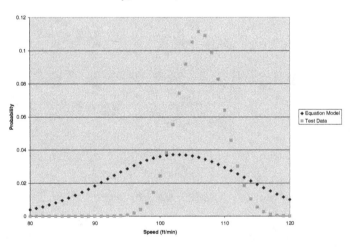

Figure M-7

Alkaline Battery, Power 8, No Sled Performance

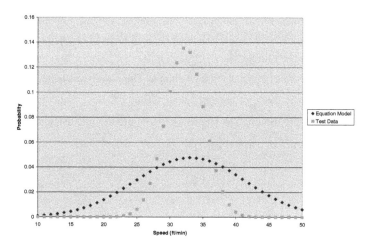

Figure M-8

Alkaline Battery, Power 8, Sled Attached Performance

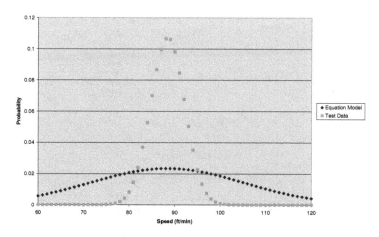

Figure M-9

Alkaline Battery, Power 5, No Sled Performance

In figure M-6, the test data may appear somewhat skewed from the prediction. However, this is not the case since this test was conducted with a freshly charged set of batteries. The data from such a sample set would be expected to fall within the upper tail of the predictive distribution, which is exactly where it falls. The data in figures M-6 through M-8 provide a reasonable match to the predictions, validating that the equation model is an accurate virtual representation of the robotic vehicle.

A different motor was obtained for the vehicle. The motor was tested to obtain the RPM as a function of torque load, battery type and power setting. A Bayesian network of the motor was constructed for the new motor. The test results were used to create the probability tables. The motor network model is shown in figure M-10.

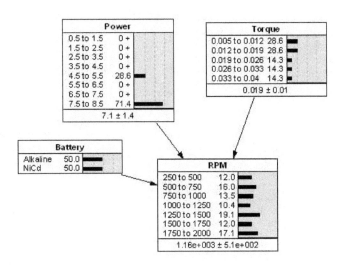

Figure M-10

Motor Bayesian Network Model

The network used a non-uniform prior for probability learning to reduce any errors from bins with no data.

The robotic vehicle was modified as shown in figure M-11.

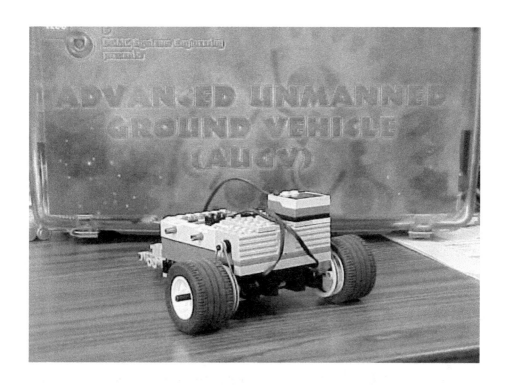

Figure M-11

Modified Robotic Vehicle

The new motors were much larger in size than the original motors. As a result, the drive train had to be changed from the gear driven system of figure M-1 to a pulley and belt drive system in figure M-11. The drive and axle pulley sizes were adjusted in the model. The engine model of figure M-3 was then removed and replaced with the Bayesian network interface blocks shown in figure M-12.

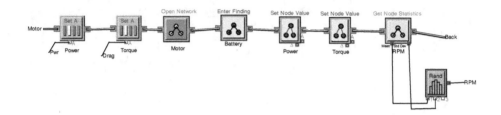

Figure M-12

Modified Robotic Vehicle Bayesian Network Model Integration

The elements of figure M-12 set the power and torque load calculated by the equation-based portions of the models as attributes and then opens and compiles the network of figure M-10. The battery type, power level and torque load variable states are passed to the network and the mean and standard deviation of the motor RPM is returned. Because the model uses a Monte Carlo method to create a probability distribution for the output prediction, the mean and standard deviation are input to a random number generator using a normal distribution of RPM.

The modified model was run to create a probability distribution prediction of the robotic vehicle with the new motor and drive train. The modified robotic car was then tested under the same conditions as figures M-6 through M-9. Multiple tests were conducted to create a probability distribution. The model correctly predicted, and test data confirmed, that the robotic vehicle could not tow the sled in the modified configuration. The other three conditions are presented in figures M-13 through M-15.

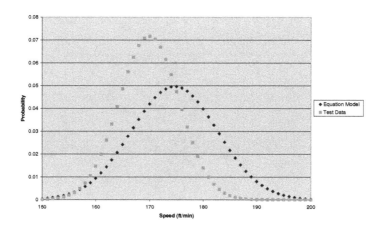

Figure M-13

Modified Vehicle, Alkaline Battery, Power 8, No Sled Performance

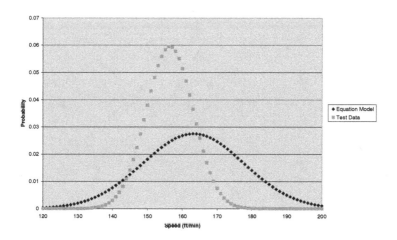

Figure M-14

Modified Vehicle, Ni-Cd Battery, Power 8, No Sled Performance

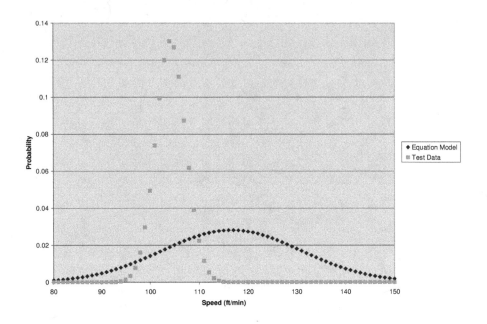

Figure M-15

Modified Vehicle, Alkaline Battery, Power 5, No Sled Performance

As can be seen in figures M-13 through M-15, the test data distribution fell within the limits of the prediction in all 3 cases, as well as correctly predicting that there was insufficient power to tow the sled.

BIBLIOGRAPHY

Acquisition Functional Working Group, (1999) "Acquisition Modeling and Simulation Comprehensive Core Body of Knowledge" Version 1.0.

Anderson, Dave and McNeil, George, (1992) "Artificial Neural Networks Technology", Technical Report for contract F30602-89-C-0082, Data & Analysis Center for Software, Rome Laboratory, Rome, NY.

Arsham, Hossein, (2002) "Systems Simulation: The Shortest Path from Learning to Applications", 7th Edition, http://ubmail.ubalt.edu/~harsham/simulation/sim.htm.

Bankes, Steve (1993) "Exploratory Modeling for Policy Analysis", *Operations Research*, Vol. 41, No. 3, May-June, pages 435-449.

Bellinger, Gene, (2002) "Modeling & Simulation: An Introduction", http://www.outsights.com/systems/modsim/modsim.htm

Blood, Ernie, (2001) "M&S at Caterpillar", Brief to the Committee on Modeling and Simulation Enhancements for 21st Century Manufacturing and Acquisition, National Research Council.

Breese, Jack and Koller, Daphne, (1997) "Bayesian Networks and Decision-Theoretic Reasoning for Artificial Intelligence", http://www.robotics.stanford.edu/~koller/BNtut/sld001.htm

Brown, CMDR David P., (1999) "Simulation Based Acquisition – Can It Live Up to Its Promises?" *Program Manager,* vol. 28, pp. 12-17.

Brown, David P., (2000) "Genetic Algorithm Solutions to Complex Games", INFT-803 Final Report, George Mason University.

Cheng, Jie and Bell, David and Liu, Weiru, (1997) "Learning Belief Networks from Data: An Information Theory Based Approach", *(proceedings of the Sixth ACM International Conference on Information and Knowledge Management, 1997)* CIKM'97

Cherkassky, Vladimir and Yunqain, Ma, (2002a) "Selecting the Loss Function for Robust Linear Regression", Submitted to *Neural Computation,* http://www.ece.umn.edu/users/cherkass/pub.html

Cherkassky, Vladimir and Yunqain, Ma, (2002b) "Multiple Model Estimation: A New Formulation for Predictive Learning", Submitted to *IEEE Transaction on Neural Networks,* http://www.ece.umn.edu/users/cherkass/pub.html

Cherkassky, Vladimir and Mulier, Filip, (1998) Learning From Data: Concepts, Theory and Methods, John Wiley and Sons, Inc. New York, NY.

Clemen, Robert T. and Really, Terrence, (1999) "Dependent Decision Analysis", August 1999.

Clemen, R.T., (1996) *Making Hard Decisions: An Introduction to Decision Analysis, 2nd Ed.*, Duxbury, Belmont, CA.

Clymer, J.R. (1999) "Simulation-Based Engineering of Complex Adaptive Systems". *Simulation*, 72 (4), 250-260.

Clymer, J.R. (1997a) "Engineering of Intelligent, Complex Adaptive Systems using an Agent-Based Architecture" in *Proceedings of the 7th Annual International Symposium of the International Council on Systems Engineering*, Los Angeles, CA.

Clymer, J.R. (1997b) "Expansionist System Design and Evaluation Methodology Based on Context Sensitive System Theory". *IEEE Transactions on Aerospace and Electronic Systems*, 33 (2), pages 686-695.

Cohen, Marvin, (1992) "Radar Fundamentals" Radar Cross Section Reduction, Georgia Institute of Technology.

Cooper, Gregory F. and Herskovits, Edward, (1992) "A Bayesian Method for the Induction of Probabilistic Networks from Data", *Machine Learning,* 9, Klewer Academic Publishers, Boston, MA, pages 309-347.

Davis, Paul K. and Zeigler, Bernard. (2000) "Multi-resolution Modeling and Integrated Families of Models". *Modeling and Simulation, Volume 9*, Appendix E, pages 1-18.
http://www.nas.edu/cpsma/nsb/mse.htm

DeClaris, John-William and Roberts, James (1997) "An Introduction to Neural Networks",
http://www.ee.umd.edu/medlab/neural/nn1.html

Doyle, John. (2000) "Virtual Engineering: Toward a Theory for Modeling and Simulation of Complex System", *Modeling and Simulation, Volume 9*, Appendix B. California Institute of Technology, pages 1-43.

DTSE&E study (1996) "Study on the Effectiveness of Modeling and Simulation in the Weapon System Acquisition Process", Final Report.

Freidman, N. and Goldszmidt, M., (1996) "Learning Bayesian Networks: The Combination of Knowledge and Statistical Data", *Machine Learning*, 20, pp. 197-243.

Friedman, N., (1998) "Learning Belief Networks in the Presence of Missing Values and Hidden Variables", *Fourteenth International Conference on Machine Learning (ICML-97)*, Vanderbilt University, Morgan Kaufmann Publishers.

Garcia, LTC Albert B., Gocke, COL Robert P. Jr., and Johnson, COL Nelson P. Jr., (1994) "Virtual Prototyping: Concept to Production", Report of the DSMC 1992-1993 Military Research Fellows, Defense Systems Management Conference Press, Fort Belvoir, VA.
Gilks, W. R., and Richardson, S., and Spiegelhalter, D. J., (1996) "Markov Chain Monte Carlo in Practice", Chapman & Hall/CRC.

Heckerman, David, Geiger, Dan, Chickering, David M., (1995) "Learning Bayesian Networks: The Combination of Knowledge and Statistical Data", *Machine Learning,* 20, Kluwer Academic Publishers, Boston, MA, pages 197-243.

Hicks & Associates, Inc., (2001) "Modeling and Simulation Survey Briefing".

Hillegas, Anne, Backschies, John, Donley, Michael, Duncan, R. Clif and Edgar, William (2001) "The Use of Modeling & Simulaton (M&S) Tools in Acquisition Program Offices: Results of a Survey", Hicks & Associates, Inc., January 31, 2001.

Hollenbach, James W. (2001) "Collaborative Achievement of Advanced Acquisition Environments", *Simulation Interoperability Workgroup publications*, Paper 01S-SIW-091, Spring, 2001.

Hollenback, Jim W. Former Director of DMSO, Interview of 11 August 2000.

Hoyt, Pam, (2000) "Independent Study on Descretizing Continuous Variables", George Mason University.

Jensen, Finn V., (1996) An Introduction to Bayesian Networks, UCL Press Limited, London, UK.

Judson, Dean H., "Bayesian Record Linkage Across Massive Databases: Problems and Prospects", U.S. Census Bureau, April 2002, http://arrowsmith2.psych.uic.edu/cci/judson/sld001.htm

Kincade, Chuck (1993) "Why RCS Reduction?" Tutorial Application of LO Technology to Naval Aircraft, McDonnell Douglas Technologies Inc.
Knight, Will, (2003) "Intel to Release Machine Learning Libraries", NewScientist.com, http://www.newscientist.com/news/news.jsp?id=ns99993691

Knott, Eugene F. and Shaeffer, John F. and Tuley, Michael T. (1985) Radar Cross Section: Its Prediction, Measurement and Reduction, Artec House, Inc., Norwood, MA.

Koller, Daphne and Pfeffer, Avi. "Object Oriented Bayesian Networks" in Geiger, D. and Shenoy, P. (eds.) *Uncertainty in Artificial Intelligence: Proceedings of the Thirteenth Conference.* Morgan Kauffman: San Francisco, pp. 302-313.

Konwin, Colonel Kenneth "Crash" and Miller, Ray (2001), "Simulation Based Acquisition: From Motivation to Implementation", *Simulation Interoperability Workgroup publications*, Paper 01S-SIW-092, Spring, 2001.

Kreft, Ita G.G., (1996) "Are Multilevel Techniques Necessary? An overview, including Simulation Studies", California State University, Los Angeles, CA, June 1996, http://www.calstatela.edu/faculty/ikreft/quarterly/quarterly.html

Larson, Harold J., (1969) Introduction to Probability Theory and Statistical Inference, John Wiley & Sons, New York.

Laskey, G. and Laskey, K.B., (2002) "Combat Identification with Bayesian Networks", *Proceedings of the Command and Control Research and Technology Symposium*, Spring 2002.

Laskey, Kathryn Blackmond (2002) SYST/STAT 664 Lecture Notes, George Mason University, http://ite.gmu.edu/~klaskey/SYST664/SYST664.html

Laskey, K.B. and Mahoney, S.M. (1997) "Network Fragments: Representing Knowledge for Constructing Probabilistic Models" in Geiger, D. and Shenoy, P. (eds.) *Uncertainty in Artificial Intelligence: Proceedings of the Thirteenth Conference*. Morgan Kauffman: San Francisco, pp. 334-341.

Laskey, K.B. and Lehner, P.E. (1994) "Metareasoning and the Problem of Small Worlds," *IEEE Transactions on Systems, Man, and Cybernetics*, 24 (11), 1643-1652.

Lerat, Jean-Philippe, (2002) "Four Applications of Bayesian Networks for Systems Engineering", *Proceedings of the 3rd European Systems Engineering Conference*, Toulouse, France, pp. 177-184.

Liu, Huan and Hussain, Farhad and Chew, Lim Tan and Dash, Manoranjan, (2002) "Discretization: An Enabling Technique", *Data Mining and Knowledge Discovery, 6*, Kluwer Academic Publishers, The Netherlands, pp. 393-423.

Maren, Alianna and Harston, Craig and Pap, Robert, (1990) Handbook of Neural Network Applications, Academic Press Inc., San Diego, CA.

Masters, George W. (1981a) Electro-Optical Systems Test and Evaluation, United States Naval Test Pilot School, NAS Patuxent River, MD.

Masters, George W. (1981b) Radar System Test and Evaluation, United States Naval Test Pilot School, NAS Patuxent River, MD.

Monti, Stefano and Cooper, Gregory F., (1997) "Learning Hybrid Bayesian Networks from Data", Technical Report ISSP-97-01, Intelligent Systems Program, University of Pittsburgh, Pittsburgh, PA.

Monti, Stefano, (1999) "Learning Hybrid Bayesian Networks from Data", Dissertation for degree of Doctor of Philosophy, University of Pittsburgh, Pittsburgh, PA.
Morgan, H. G., Henrion, M. and Small, M. (1990) A Guide to Dealing With Uncertainty in Quantitative Risk and Policy Analysis, Cambridge University Press.

Murphy, K., (2000) "A Brief Introduction to Graphical Models and Bayesian Networks", http://www.cs.berkeley.edu/~murphyk/Bayes/bayes.html

Myers, James William (1999) "Stochastic Algorithms for Learning with Incomplete Data: An Application to Bayesian Networks" George Mason University.

Niedermayer, Daryle, (1998) "An Introduction to Bayesian Networks and their Contemporary Applications", http://www.gpfn.sk.ca/~daryle/papers/bayesian_networks/bayes.html

Palmore, Julian. (1993) "Variable Resolution Modeling in Mathematics," *Proceedings of Conference on Variable Resolution Modeling*, Davis and Hillestad (eds.), RAND, CF-103, Santa Monica, Calif.

Pew, Richard and Mavor, Anne. (1998) "Modeling Human and Organizational Behavior: Applications to Military Simulations", National Academy Press, Washington DC.

Reynolds, Paul F., Natrajan, Anand, and Srinivasan, Sudhir, (1997) "Consistency Maintenance in Multiresolution Simulation", *ACM Transactions on Modeling and Computer Simulation (TOMACS),* Volume 7, Issue 3, July 1997, pp. 368-392.

Ryan, P. A. (1992) "Computer Models for RCS Prediction", Radar Cross Section Reduction, Georgia Institute of Technology.

Sahami, M., Dumais, D., Heckerman D., and Horvitz, E. "A Bayesian Approach to Filtering Junk E-mail", *AAAI'98 Workshop on Learning for Text Categorization*, Madison, WI, July 1998.

Scheckeler, Günther. 1993. "A Hierarchy of Models at Different Resolution Levels" in Davis, P.K. and Hillestad, R. (1993) (eds.) *Proceedings of Conference on Variable Resolution Modeling*, RAND.

Shaeffer, John F. (1992) "High Frequency Scattering Mechanisms", Radar Cross Section Reduction, Georgia Institute of Technology.

Smith, Leslie, (2001) "An Introduction to Neural Networks", Centre for Cognitive and Computational Neuroscience, Department of Computing and Mathematics, University of Stirling, UK, http://www.cs.stir.ac.uk/~lss/NNIntro/InvSlides.html

Smith, Roger D. (1999) "Multi-Resolution Modeling". Military Techniques and Technology, Distributed Simulation Technology, Inc., pages 10-1 to 10-31.

Stimson, George W. (1983) <u>Introduction to Airborne Radar</u>, Hughes Aircraft Co., El Segundo, CA.

Stutz, J., Taylor, W., and Cheeseman, P., (1998) "AutoClass C – General Information" NASA Ames Research Center, http://ic-www.arc.nasa.gov/ic/projects/bayes-group/autoclass/autoclass-c-program.html#AutoClass_C

Xu, Ning, (2003) "A Comparison of Discretization Methods for Bayesian Networks", Technical Report, Systems Engineering and Research Department, George Mason University, Fairfax, VA.

Yin, Robert K. (1994) <u>Case Study Research: Design and Methods</u>, Second Edition, Sage Publications Inc., Thousand Oaks, CA.

Zeigler, Bernard. (1990) "Object-oriented Simulation with Hierarchical, Modular Models: Intelligent Agents and Endomorphic Systems", Academic Press, Boston, Mass., and San Diego, Calif.

Zeigler, Bernard, H. Praehofer, and Jerry Rozenblit. (1993) "Integrating System Formalisms: How Object Orientation Supports CAST for Intelligence Systems Design". *Journal of Systems Engineering*, 3, 209-219.

Zhong, Shi and Cherkassky, Vladimir, (2000) "Image Denoising using Wavelet Thresholding and Model Selection", *Proc. IEEE Int. Conf. on Image processing,* Vancouver, BC, Canada.

Zhong, Shi and Cherkassky, Vladimir, (1999) "Factors Controlling Generalization Ability of MLP Networks", *Proc. IEEE Int. Joint Conf. on Neural Networks*, Washington D.C.

Zittel, Randy C., (1998) "Simulation, IPPD & Systems Engineering: To The Next Level", *Proceedings of the 1998 Symposium of the International Council on Systems Engineering*, Vancouver, B.C. Canada.

Wissenschaftlicher Buchverlag bietet

kostenfreie

Publikation

von

wissenschaftlichen Arbeiten

Diplomarbeiten, Magisterarbeiten, Master und Bachelor Theses
sowie Dissertationen, Habilitationen und wissenschaftliche Monographien

Sie verfügen über eine wissenschaftliche Abschlußarbeit zu aktuellen oder zeitlosen
Fragestellungen, die hohen inhaltlichen und formalen Ansprüchen genügt,
und haben **Interesse an einer honorarvergüteten Publikation**?

Dann senden Sie bitte erste Informationen über Ihre Arbeit per Email
an info@vdm-verlag.de. Unser Außenlektorat meldet sich umgehend bei Ihnen.

VDM Verlag Dr. Müller Aktiengesellschaft & Co. KG
Dudweiler Landstraße 125a
D - 66123 Saarbrücken

www.vdm-verlag.de

www.ingramcontent.com/pod-product-compliance
Lightning Source LLC
La Vergne TN
LVHW022303060326
832902LV00020B/3250